Cover photos, clockwise from top left:
Women selling caterpillars in Bangui, Central African Republic (P. Vantomme)
Gold-painted crickets on top of Belgian chocolates (P. Vantomme)
Black soldier fly in a mass-rearing unit (L. Heaton)
Appetizers prepared with insects (T. Calame)
Coleoptera species used as a food colorant (A. Halloran)
Palm weevil larvae (O. Ndoye)

Edible insects: future prospects for food and feed security

FAO
FORESTRY
PAPER

171

by
Arnold van Huis
Joost Van Itterbeeck
Harmke Klunder
Esther Mertens
Afton Halloran
Giulia Muir
and
Paul Vantomme

FOOD AND AGRICULTURE ORGANIZATION OF THE UNITED NATIONS
Rome, 2013

ISBN 978-92-5-107595-1 (print)
E-ISBN 978-92-5-107596-8 (PDF)

Contents

Boxes

Figures

Tables

Foreword

It is widely accepted that by 2050 the world will host 9 billion people. To accomodate this number, current food production will need to almost double. Land is scarce and expanding the area devoted to farming is rarely a viable or sustainable option. Oceans are overfished and climate change and related water shortages could have profound implications for food production. To meet the food and nutrition challenges of today – there are nearly 1 billion chronically hungry people worldwide – and tomorrow, what we eat and how we produce it needs to be re-evaluated. Inefficiencies need to be rectified and food waste reduced. We need to find new ways of growing food.

Edible insects have always been a part of human diets, but in some societies there is a degree of distaste for their consumption. Although the majority of edible insects are gathered from forest habitats, innovation in mass-rearing systems has begun in many countries. Insects offer a significant opportunity to merge traditional knowledge and modern science in both developed and developing countries.

This publication has its beginnings in an effort in FAO's Forestry Department to recognize the traditional practices of gathering insects for food and income, and to document the related ecological impacts on forest habitats. Thereafter, FAO embraced the opportunity to collaborate with the Laboratory of Entomology at Wageningen University in the Netherlands – an institution at the forefront of fundamental and applied research on insects as food and feed. This combined effort has since gained momentum and is unfolding into a broad-based effort at FAO to examine the multiple dimensions of insect gathering and rearing as a viable option for alleviating food insecurity.

This book draws on a wide range of scientific research on the contribution that insects make to ecosystems, diets, food security and livelihoods in both developed and developing countries. We hope that it will help raise the profile of insects as sources of food and feed in national and international food agencies. We also hope that it attracts the attention of farmers, the media, the public at large and decision-makers in governments, multilateral and bilateral donor agencies, investment firms, research centres, aid agencies and the food and feed industries. Above all, it is our hope that this publication will raise awareness of the many valuable roles that insects play in sustaining nature and human life and will also serve to document the contribution insects already make to diversifying diets and improving food security.

Eduardo Rojas-Briales
Assistant Director-General
FAO Forestry Department

Ernst van den Ende
Managing Director
Department of Plant Sciences Group
Wageningen University and Research Centre

Abbreviations

BCE	Before Common Era
BSE	bovine spongiform encephalopathy
CABIN	Central African Biodiversity Information Network
CE	Common Era
CGRFA	FAO's Commission on Genetic Resources for Food and Agriculture
CH_4	methane
CO_2	carbon dioxide
CRGB	Centre de Recherche pour la Gestion de la Biodiversité (Benin)
EFSA	European Food Safety Agency
ESBL	extended spectrum beta-lactamase
EU	European Union
FBF	fortified blended foods
g	gram
GHG	greenhouse gas
GWP	global warming potential
HACCP	Hazard Analysis Critical Control Points system
IFIF	International Feed Industry Federation
INFOODS	International Network of Food Data Systems
IPM	integrated pest management
kg	kilogram
N_2O	nitrous oxide
NGO	non-governmental organization
NWFP	non-wood forest product
PAP	processed animal protein
RDA	recommended dietary allowances
SEPALI	Madagascar Organization of Silk Workers
SPS Agreement	Agreement on the Application of Sanitary and Phytosanitary Measures
VENIK	Dutch Insect Farmers Association
WHO	World Health Organization
WTO	World Trade Organization
WUR	Wageningen University and Research Centre

Authors' preface

Insects are often considered a nuisance to human beings and mere pests for crops and animals. Yet this is far from the truth. Insects provide food at low environmental cost, contribute positively to livelihoods, and play a fundamental role in nature. However, these benefits are largely unknown to the public. Contrary to popular belief, insects are not merely "famine foods" eaten in times of food scarcity or when purchasing and harvesting "conventional foods" becomes difficult; many people around the world eat insects out of choice, largely because of the palatability of the insects and their established place in local food cultures.

In 2008, within the framework of the Wageningen University–FAO partnership, a few researchers came together and began reviewing an extensive array of published and unpublished research and information on insect rearing and consumption. Their intention was to break down the aforementioned misconceptions and contribute positively to the development of the edible insects sector. The subject of edible insects inherently covers a wide range of thematic areas, from the conservation of habitats where insects are harvested to insect ecology, the artificial rearing of insect species, the processing of insects into food and feed products, and the labelling and marketing of insect-based food and feed products. This publication, therefore, draws from a wide range of disciplines and areas of expertise. It is a multidisciplinary effort involving technical experts specializing in forestry, animal farming, nutrition, the feed industry, legislation and food security policies.

This publication marks the first attempt by FAO to document all aspects of the insect food and feed value chain, with the aim of enabling a comprehensive assessment of the contribution of insects to food and feed security. It includes original research from around the world, such as that carried out at Wageningen University. It also incorporates findings from the International Expert Consultation on Assessing the Potential of Insects as Food and Feed in Assuring Food Security, which took place at FAO headquarters in Rome, Italy, on 23–25 January 2012. This meeting marked the beginning of a dialogue between agricultural experts from various backgrounds and fostered an exchange of information on the potential benefits of using insects for food and feed as part of a broader strategy to achieve global food security. The participants at this meeting provided the authors with a wealth of supplementary data and valuable insights. These helped to shape the form and content of this book and its conclusions, which it is hoped can provide a basis for solutions to alleviate food insecurity.

Insect rearing for food and feed remains a sector in its infancy, and key future challenges will likely emerge as the field evolves. As such, readers are encouraged to contact the authors with feedback on this book. Such contributions will undoubtedly assist the future development of the sector.

Since the science of edible insects is still at a relatively pioneering stage, it boasts only a few scientists of renown. One of those, Gene R. DeFoliart (1925–2013), died shortly before this book was published. He spent his long academic career passionately raising awareness of insects as a global food source, and he continued his work in this area long after his retirement in July 1991. He was also the founder of *The Food Insects Newsletter.* The authors dedicate this book to his memory.

Acknowledgements

This book was made possible by the valuable contributions of many people with a variety of backgrounds and from many parts of the world. Their ideas, papers and professional activities all played a fundamental role in shaping this publication. Among them, special thanks are extended to the 75 participants at the Expert Consultation Meeting on Assessing the Potential of Insects as Food and Feed in Assuring Food Security, which was held in Rome on 23–25 January 2012.[1] Special thanks go to those who reviewed specific chapters of this book: Christian Borgemeister, Eraldo Medeiros Costa-Neto, David Drew, Florence Dunkel, Jørgen Eilenberg, Ying Feng, Parimalendu Haldar, Yupa Hanboonsong, Antoine Hubert, Annette Bruun Jensen, Nonaka Kenichi, Andrew Müller, Maurizio Paoletti, Julieta Ramos Elorduy Blásquez, Nanna Roos, Oliver Schneider, Severin Tchibozo and Alan L. Yen.

Several FAO staff members kindly volunteered to review chapters relevant to their areas of expertise: Philippe Ankers, Jan Breithaupt, Carmen Bullón, Ruth Charrondiere, Persijn Diedelinde, Patrick Durst, Graham Hamley, Martin Hilmi, Edgar Kaeslin, Blaise Kuemlangan, Harinder Makar, Verena Nowak, Koroma Suffyan, Patrice Talla, Pieter Van Lierop and Philine Wehling. We thank them for their willingness to contribute to such an interdisciplinary effort. Special thanks also go to staff at Wageningen University, including Sarah van Broekhoven and Dennis Oonincx.

The authors are grateful to David McDonald and Alastair Sarre for editing, Yde Jongema for checking the Latin names of the insects, Kate Ferrucci for design and layout, Susy Tafuro and Lucia Travertino Grande for the administrative handling of the manuscript from printing through to distribution, and Maria DiCristofaro and Alison Small for media support. Special thanks are due to Eva Müller, the Director of the Forest Economics, Policy and Products Division at FAO, and Michael Martin, the previous director, who both fully supported the Edible Insects Programme.

Above all, the authors acknowledge all people around the world for whom eating insects is and has always been an integral part of daily life. They have provided time-honoured understandings of edible insects and remain custodians of valuable knowledge on the important roles that insects play in daily lives. These peoples are a key to the continued practice of eating insects and the potential of edible insects as future sources of food and feed.

[1] For futher information, see www.fao.org/forestry/edibleinsects/74848/en/.

Executive summary

This book assesses the potential of insects as food and feed and gathers existing information and research on edible insects. The assessment is based on the most recent and complete data available from various sources and experts around the world.

Insects as food and feed emerge as an especially relevant issue in the twenty-first century due to the rising cost of animal protein, food and feed insecurity, environmental pressures, population growth and increasing demand for protein among the middle classes. Thus, alternative solutions to conventional livestock and feed sources urgently need to be found. The consumption of insects, or **entomophagy**, therefore contributes positively to the environment and to health and livelihoods.

This publication grew from a small effort in 2003 in the FAO Forestry Department to document the role of insects in traditional livelihood practices in Central Africa and to assess the impact of harvesting insects in their natural habitats on the sustainability of forests. This effort has since unfolded into a broad-based effort to examine the multiple dimensions of insect gathering and rearing to clarify the potential that insects offer for improving food security worldwide. The purpose of this book is to bring together for the first time the many opportunities for, and constraints on, using insects as food and feed.

THE ROLE OF INSECTS

It is estimated that insects form part of the traditional diets of at least 2 billion people. More than 1 900 species have reportedly been used as food. Insects deliver a host of ecological services that are fundamental to the survival of humankind. They also play an important role as pollinators in plant reproduction, in improving soil fertility through waste bioconversion, and in natural biocontrol for harmful pest species, and they provide a variety of valuable products for humans such as honey and silk and medical applications such as maggot therapy. In addition, insects have assumed their place in human cultures as collection items and ornaments and in movies, visual arts and literature. Globally, the most commonly consumed insects are beetles (Coleoptera) (31 percent), caterpillars (Lepidoptera) (18 percent) and bees, wasps and ants (Hymenoptera) (14 percent). Following these are grasshoppers, locusts and crickets (Orthoptera) (13 percent), cicadas, leafhoppers, planthoppers, scale insects and true bugs (Hemiptera) (10 percent), termites (Isoptera) (3 percent), dragonflies (Odonata) (3 percent), flies (Diptera) (2 percent) and other orders (5 percent).

CULTURE

Entomophagy is heavily influenced by cultural and religious practices, and insects are commonly consumed as a food source in many regions of the world. In most Western countries, however, people view entomophagy with disgust and associate eating insects with primitive behaviour. This attitude has resulted in the neglect of insects in agricultural research. Despite historical references to the use of insects for food, the topic of entomophagy has only very recently started to capture public attention worldwide.

INSECTS AS A NATURAL RESOURCE

Edible insects inhabit a large variety of habitats, from aquatic ecosystems and farmed land to forests. Until recently, insects were a seemingly inexhaustible resource obtainable by harvesting from nature. However, some edible insect species are now in peril. A number of anthropogenic factors, such as overharvesting, pollution, wildfire and habitat degradation, have contributed to a decline in many edible insect populations. Climate change will likely affect the distribution and availability of edible insects in ways that are still relatively unknown. This publication includes case studies from several regions on the conservation strategies and semi-cultivation practices of rural people to protect insect species and their host plants. Such efforts contribute to improved habitat conservation.

ENVIRONMENTAL OPPORTUNITIES

The environmental benefits of rearing insects for food and feed are founded on the high feed conversion efficiency of insects. Crickets, for example, require only 2 kilograms of feed for every 1 kilogram of bodyweight gain. In addition, insects can be reared on organic side-streams (including human and animal waste) and can help reduce environmental contamination. Insects are reported to emit fewer greenhouse gases and less ammonia than cattle or pigs, and they require significantly less land and water than cattle rearing. Compared with mammals and birds, insects may also pose less risk of transmitting zoonotic infections to humans, livestock and wildlife, although this topic requires further research.

NUTRITION FOR HUMAN CONSUMPTION

Insects are a highly nutritious and healthy food source with high fat, protein, vitamin, fibre and mineral content. The nutritional value of edible insects is highly variable because of the wide range of edible insect species. Even within the same group of species, nutritional value may differ depending on the metamorphic stage of the insect, the habitat in which it lives, and its diet. For example, the composition of unsaturated omega-3 and six fatty acids in mealworms is comparable with that in fish (and higher than in cattle and pigs), and the protein, vitamin and mineral content of mealworms is similar to that in fish and meat.

FARMING SYSTEMS

Most edible insects are harvested in the wild. However, some insect species, such as bees and silkworms, have a long history of domestication because of the value of their products. Insects are also reared in large numbers for the purposes of biological control (e.g. as predators and parasitoids), health (e.g. maggot therapy) and pollination. The concept of farming insects for food is, however, relatively new; an example of rearing insects for human consumption in the tropics is cricket farming in the Lao People's Democratic Republic, Thailand and Viet Nam.

In temperate zones, insect farming is performed largely by family-run enterprises that rear insects such as mealworms, crickets and grasshoppers in large quantities, mainly as pets or for zoos. Some of these firms have only recently been able to commercialize insects as food and feed, and the part of their production intended for direct human consumption is still minimal.

A few industrial-scale enterprises are in various stages of start-up for rearing mass quantities of insects such as black soldier flies. They are mainly for consumption as whole insects or to be processed into meal for feed. Critical elements for successful rearing include research on biology, rearing condition control and diet formulas for the farmed insect species. Current production systems are expensive, with many patents pending. A major challenge of such industrial-scale rearing is the development of automation processes to make plants economically competitive with the production of meat (or meat-substitutes like soy) from traditional livestock or farming sources.

INSECTS AS ANIMAL FEED

Recent high demand and consequent high prices for fishmeal/soy, together with increasing aquacultural production, is pushing new research into the development of insect protein for aquaculture and poultry. Insect-based feed products could have a similar market to fishmeal and soy, which are presently the major components used in feed formulae for aquaculture and livestock. Available evidence suggests that insect-based feeds are comparable with fishmeal and soy-based feed formulae. Live and dead insects already have established niche markets, mainly as feed given to pets and at zoos.

PROCESSING

Insects are often consumed whole but can also be processed into granular or paste forms. Extracting proteins, fats, chitin, minerals and vitamins is also possible. At present, such extraction processes are too costly and will need to be further developed to render them profitable and applicable for industrial use in the food and feed sectors.

FOOD SAFETY AND PRESERVATION

The processing and storage of insects and their products should follow the same health and sanitation regulations as for any other traditional food or feed items in order to ensure food safety. Because of their biological makeup, several issues should be considered, such as microbial safety, toxicity, palatability and the presence of inorganic compounds. Specific health implications should also be considered when insects for feed are reared on waste products such as manure or slaughterhouse waste. Evidence of allergies induced through the ingestion of insects is scarce, but does exist. Some cases have been reported of allergic reactions to arthropods.

LIVELIHOOD IMPROVEMENT

Insect gathering and rearing as minilivestock at the household level or industrial scale can offer important livelihood opportunities for people in both developing and developed countries. In developing countries, some of the poorest members of society, such as women and landless dwellers in urban and rural areas, can easily become involved in the gathering, cultivation, processing and sale of insects. These activities can directly improve their own diets and provide cash income through the selling of excess production as street foods. Insects can be directly and easily collected from nature or farmed with minimal technical or capital expenditure (i.e. for basic harvesting/rearing equipment). Rearing insects may also require minimal land or market introduction efforts, as insects already form part of some local food cultures.

Protein and other nutritional deficiencies are typically more widespread in disadvantaged segments of society and during times of social conflict and natural disaster. Because of their nutritional composition, accessibility, simple rearing techniques and quick growth rates, insects can offer a cheap and efficient opportunity to counter nutritional insecurity by providing emergency food and by improving livelihoods and the quality of traditional diets among vulnerable people.

ECONOMIC DEVELOPMENT

Gathering and farming insects can offer employment and cash income, either at the household level or in larger, industrial-scale operations. In developing countries in Southern and Central Africa and Southeast Asia, where demand for edible insects exists and where it is relatively easy to bring insects to market, the process of insect gathering, rearing and processing into street foods or for sale as chicken and fish feed is easily within reach of small-scale enterprises. With only a few exceptions, international trade in insects for food is insignificant. The trade that does exist to developed countries is often driven by demand from immigrant communities or because of the development of niche markets that sell exotic foods. Border trade in edible insects is significant, mainly in Southeast Asia and Central Africa.

COMMUNICATION

The polarity of views surrounding the practice of entomophagy necessarily requires tailormade communication approaches for each of the various stakeholders. In the tropics, where entomophagy is well established, media communication strategies should promote edible insects as valuable sources of nutrition to counter the growing westernization of diets. Western societies require tailored media communication strategies and educational programmes that address the disgust factor. Influencing the public at large as well as policymakers and investors in the food and feed sectors by providing validated information on the potential of insects as food and feed sources can help to push insects higher on political, investment and research agendas worldwide.

LEGISLATION

Regulatory frameworks governing food and feed chains have expanded tremendously in the last 20 years; however, regulations governing insects as food and feed sources are still largely absent. For developed countries, the absence of clear legislation and norms guiding the use of insects as food and feed is among the major limiting factors hindering the industrial development of farming insects to supply the food and feed sectors. In developing countries, the use of insects for human or animal food is, in practice, more tolerated then regulated. The feed sector seems to take the lead in pushing for the development of more insect-encompassing norms, while the "novel food" concept seems to be emerging as a leading instrument for setting rules and standards for the use of insects in human foods.

THE WAY FORWARD

Any effort to release the huge potential that insects offer for enhancing food security requires that the following four key bottlenecks and challenges are addressed simultaneously. First, further documentation is needed on the nutritional values of insects in order to more efficiently promote insects as healthy food. Second, the environmental impacts of harvesting and farming insects must be investigated to enable comparison with traditional farming and livestock rearing practices that may be more environmentally damaging. Third, clarification and augmentation of the socio-economic benefits that insect gathering and farming can offer is needed, in particular to enhance the food security of the poorest of society. Finally, a clear and comprehensive legal framework at (inter)national levels is needed to pave the way for more investment, leading to the full development (from the household to the industrial scale) of production and international trade in insect products as food and feed sources.

1. Introduction

The practice of eating insects (Box 1.1) is known as **entomophagy**. Many animals, such as spiders, lizards and birds, are entomophagous, as are many insects. People throughout the world have been eating insects as a regular part of their diets for millennia. Although this practice should be specified as *human* entomophagy, throughout this book entomophagy refers to human entomophagy. The earliest citing of entomophagy can be found in biblical literature; nevertheless, eating insects was, and still is, taboo in many westernized societies. The unconventional nature of entomophagy has meant that farming insects for food and feed has largely been absent from the great agricultural innovations in livestock farming that emerged in past centuries – with a few exceptions, such as bees, silkworms and scale insects (from which a red colorant is derived). Insects have also failed to feature on the agendas of agricultural research and development agencies worldwide, including at FAO. Until recently, references to insects for food and feed have been largely anecdotal. It is therefore unsurprising that insects are still lacking from the diets of many rich nations and that their sale for human consumption remains part of a niche food sector of novelty snacks.

Nevertheless, insect consumption is not a new concept in many parts of the world. From ants to beetle larvae – eaten by tribes in Africa and Australia as part of their subsistence diets – to the popular, crispy-fried locusts and beetles enjoyed in Thailand, it is estimated that insect-eating is practised regularly by at least 2 billion people worldwide. More than 1 900 insect species have been documented in literature as edible, most of them in tropical countries. The most commonly eaten insect groups are beetles, caterpillars, bees, wasps, ants, grasshoppers, locusts, crickets, cicadas, leaf and planthoppers, scale insects and true bugs, termites, dragonflies and flies.

BOX 1.1

What are insects?

The word insect derives from the Latin word *insectum*, meaning "with a notched or divided body", literally "cut into sections", from the fact that insects' bodies have three parts. Pliny the Elder created the word, translating the Greek word ἔντομος (entomos) or insect (as in entomology, which was Aristotle's term for this class of life), also in reference to their "notched" bodies. The term was first documented in English in 1601 in Holland's translation of Pliny (Harpe and McCormack, 2001).

Insects are a class of animals within the arthropod group that have a chitinous exoskeleton, a three-part body (head, thorax and abdomen), three pairs of jointed legs, compound eyes and two antennae. They are among the most diverse groups of animals on the planet: there are more than 1 million described species, which is more than half of all known living organisms. The total number of species is estimated at 6–10 million, and the class potentially represents over 90 percent of the differing animal life forms on Earth. Insects may be found in nearly all environments, although only a small number of species occur in the oceans, a habitat dominated by another arthropod group, the crustaceans.

Insect facts:

- Insects have an exoskeleton to protect them from the environment.
- Insects are the only winged invertebrates.

Continues

Box 1.1 continued

- Insects are cold-blooded.
- Insects undergo metamorphosis to be able to adapt to seasonal variations.
- Insects reproduce quickly and have large populations.
- Insects' respiratory systems – networks of tracheal tubes – are tolerant of air and vacuum pressure, high-altitude flight and radiation.
- Insects often do not need parental care.

Source: Delong, 1960.

This publication also covers other arthropod species eaten by humans, such as spiders and scorpions, which, taxonomically speaking, are not insects.

1.1 WHY EAT INSECTS?

Overall, entomophagy can be promoted for three reasons:

- Health:
  - Insects are healthy, nutritious alternatives to mainstream staples such as chicken, pork, beef and even fish (from ocean catch).
  - Many insects are rich in protein and good fats and high in calcium, iron and zinc.
  - Insects already form a traditional part of many regional and national diets.
- Environmental:
  - Insects promoted as food emit considerably fewer greenhouse gases (GHGs) than most livestock (methane, for instance, is produced by only a few insect groups, such as termites and cockroaches).
  - Insect rearing is not necessarily a land-based activity and does not require landclearing to expand production. Feed is the major requirement for land.
  - The ammonia emissions associated with insect rearing are also far lower than those linked to conventional livestock, such as pigs.
  - Because they are cold-blooded, insects are very efficient at converting feed into protein (crickets, for example, need 12 times less feed than cattle, four times less feed than sheep, and half as much feed as pigs and broiler chickens to produce the same amount of protein).
  - Insects can be fed on organic waste streams.
- Livelihoods (economic and social factors):
  - Insect harvesting/rearing is a low-tech, low-capital investment option that offers entry even to the poorest sections of society, such as women and the landless.
  - Minilivestock offer livelihood opportunities for both urban and rural people.
  - Insect rearing can be low-tech or very sophisticated, depending on the level of investment.

1.2 WHY FAO?

Since 2003, FAO has been working on topics pertaining to edible insects in many countries worldwide. FAO's contributions cover the following thematic areas:

- generation and sharing of knowledge through publications, expert meetings and a webportal on edible insects;
- awareness-raising on the role of insects with the general public through media collaboration (e.g. newspapers, magazines and TV);
- support to member countries through field projects (e.g. the Laos Technical Cooperation Project);
- networking and multidisciplinary interactions (e.g. stakeholders working with nutrition, feed and legislation-related issues) with various sectors within and outside FAO.

Some of the most important milestones are listed below.

1.2.1 Central African study documenting the role of caterpillars

The Non-Wood Forest Products Programme of the FAO Forestry Department initiated a review in 2003 to describe the contribution of edible insects to diets in Central Africa. Four case studies and many other studies were commissioned in Central Africa with an emphasis on the Congo Basin because of the significant consumption of wild insects from important forestry resources and wildlife ecosystems. The report *Contribution of Forest Insects to Food Security: the Example of Caterpillars in Central Africa* quantified the role of edible insects as food and, as such, initiated a discussion on entomophagy as an important practice in food security. The summary and conclusions of the publication were taken up in the Overseas Development Institute's Wildlife Policy Briefing, which helped to further raise awareness among decision-makers in forestry and in bushmeat crisis discussions of the important role that edible insects play in the food security of forest-dependent people.

1.2.2 Conference in Chiang Mai, Thailand

In February 2008, the FAO Regional Office for Asia and the Pacific organized an international workshop in Chiang Mai, Thailand, titled *Forest Insects as Food: Humans Bite Back*. The workshop brought together many world experts on entomophagy, focusing specifically on the science, management, collection, harvesting, processing, marketing and consumption of edible forest insects, as well as their potential to be reared commercially by local farmers. The proceedings of the Chiang Mai workshop aimed to raise awareness of the potential of edible forest insects as a food source, document the contribution of edible insects to rural livelihoods, and highlight linkages to sustainable forest management and conservation.

1.2.3 Laos Technical Cooperation Programme, 2010–2013

FAO is implementing a technical cooperation project in the Lao People's Democratic Republic in the period 2010–2013, called "Sustainable insect farming and harvesting for better nutrition, improved food security, and household income generation". The project is an immediate response to several interventions identified in the Lao People's Democratic Republic National Nutrition Strategy and National Plan of Action for Nutrition, which was finalized and endorsed in December 2009, namely improving nutrient intake and addressing underlying causes (through improved access to food and increasing and diversifying domestic food production).

The project is focusing on strengthening the existing role of insects as complementary food in local diets, recognizing the role of traditional insect-collecting in the wild, by enhancing the sustainability, safety and efficiency of insect collection, preparation, post-harvest processing and consumption, as well as expanding insect farming.

1.2.4 FAO–WUR collaboration

As a follow up to the FAO workshop in Chiang Mai in 2008, the Non-Wood Forest Products Programme of the FAO Forestry Department and the Wageningen University and Research Centre (WUR) (Laboratory of Entomology) initiated a collaborative effort to promote entomophagy. The first step was the creation of a policy note, *Promoting the Contribution of Edible Forest Insects in Assuring Food Security*, for the FAO Forestry Department. This note outlined the long-term strategies for FAO in making the edible insects programme an integrated, regular FAO programme and creating awareness among (inter)national organizations/agencies and donors dealing with food security. In 2010, two researchers and authors of the present publication, Arnold van Huis and Joost Van Itterbeeck from WUR, worked at FAO for several months. A bibliography of publications concerning edible insects and a database on "who is who" in the entomophagy world was developed based on a widely disseminated questionnaire. Additionally, the writing of the present publication commenced, along with preparations for conceptualizing and holding an international expert consultation in January 2012.

1.2.5 Expert consultation meeting

The Expert Consultation Meeting on Assessing the Potential of Insects as Food and Feed in Assuring Food Security took place on 23–25 January 2012 at FAO headquarters in Rome. Jointly organized by FAO and WUR with the financial support of the Government of the Netherlands, the meeting aimed to open a dialogue and foster an exchange of information and expertise on the potential benefits of using insects for food and feed as part of a broader strategy to achieve global food security. Fifty-seven experts from international agencies, scientific institutions and private-sector stakeholders, together with staff from relevant FAO disciplines (nutrition, aquaculture, livestock, veterinary science, food safety, forestry and conservation) attended the meeting. These experts and entrepreneurs – specialized in various aspects of insect rearing, plant protection and food engineering – mapped the current state of the art and identified knowledge gaps in the following thematic areas: insect ecology and biology; farming insects; insects as livestock and fish feed; nutrition; processing and trade; food and feed safety; communication strategies; and policies to achieve food security.

1.2.6. Webportal of edible insects

FAO has maintained a webportal on edible insects since 2010. It provides basic information on the use and potential of edible insects as well as relevant weblinks, such as to the proceedings of the 2008 Chiang Mai workshop, information on the Expert Consultation Meeting in Rome in 2012, and other relevant technical information, videos and media coverage. The address of the webportal is www.fao.org/forestry/edibleinsects.

2. The role of insects

2.1 BENEFICIAL ROLE OF INSECTS FOR NATURE AND HUMANS

Over the past 400 million years, evolution has produced a wide variety of arthropod species adapted to their environments. About 1 million of the 1.4 million described animal species on earth are insects, and millions more are believed to exist. Contrary to popular belief, of the 1 million described insect species, only 5 000 can be considered harmful to crops, livestock or human beings (Van Lenteren, 2006).

2.1.1 Benefits for nature

Insects deliver a host of ecological services fundamental to the survival of humankind. For instance, insects play an important role in **plant reproduction**. An estimated 100 000 pollinator species have been identified and almost all of these (98 percent) are insects (Ingram, Nabhan and Buchmann, 1996). Over 90 percent of the 250 000 flowering plant species depend on pollinators. This is also true for three-quarters of the 100 crop species that generate most of the world's food (Ingram, Nabhan and Buchmann, 1996). Domesticated bees alone pollinate an estimated 15 percent of these species. The importance of this ecological service for agriculture and nature more generally is undisputed.

Insects play an equally vital role in **waste biodegradation**. Beetle larvae, flies, ants and termites clean up dead plant matter, breaking down organic matter until it is fit to be consumed by fungi and bacteria. In this way, the minerals and nutrients of dead organisms become readily available in the soil for uptake by plants. Animal carcasses, for example, are consumed by fly maggots and beetle larvae. Dung beetles – of which there are about 4 000 known species – also play a significant role in decomposing manure. They can colonize a dung heap within 24 hours, preventing flies from developing on them. If the dung remains on the soil surface, about 80 percent of the nitrogen is lost to the atmosphere; the presence of dung beetles, however, means carbon and minerals are recycled back to the soil, where they further decompose as humus for plants. When cattle were introduced to Australia in 1788, waste biodegradation became an immediate problem, as endemic dung beetles were simply insufficient to decompose the increased amounts of manure. Australian dung beetles had adapted to the dung of marsupials (e.g. kangaroos), which differs from bovine dung in various ways, including in size, texture and water content (Bornemissza, 1976). The Australian Dung Beetle Project was initiated to solve the problem, and dung beetles were introduced to the continent from South Africa, Europe and Hawaii (of 46 introduced species, 23 established).

BOX 2.1

Outbreaks of the brown planthopper

Brown planthoppers (*Nilaparvata lugens*) cause considerable damage by sucking sap from rice plants, causing them to wilt and die. They also transmit three viral diseases that stunt rice plants and prevent grain formation. When using pesticides, beneficial insects that prey on planthoppers are killed when insecticides are used injudiciously. Beneficial insects that feed on planthoppers keep pest populations below outbreak levels. However, when this balance is disrupted, planthopper outbreaks occur.

Beneficial fauna, including insects, buttress the natural resistance of agro-ecosystems. Pest insects have a large array of natural enemies, predators and parasitoids, keeping them under economic threshold levels. However, by using insecticides, vulnerable beneficial insects can be killed quicker than the targeted pest insect. One reason for this is that the target pest is often better protected (such as stem borers by the stem and mites by webs) than beneficial insects, which need to forage. Following the application of a synthetic pesticide, the population of the pest insect first decreases but then increases exponentially, because the pest insect can now develop without being constrained by attacks from beneficial insects. A notorious example of this is the outbreak of the brown planthopper in rice instigated by the use of pesticides (Box 2.1) (Heinrichs and Mochida, 1984).

Virtually all agro-ecosytems benefit from insects because they can **naturally control harmful pest species**. The number of insects that parasitize or prey on other insects is vast. Ten percent of all insects are parasitoids (Godfray, 1994). Entire orders of insects – such as Odonota (dragonflies) and Neuroptera (net-winged insects such as lacewings and antlions) – are predators. A large percentage of true bugs (Hemiptera), beetles (Coleoptera), flies (Diptera) and wasps, bees and ants (Hymenoptera) are also predators. The number of beneficial insect species in the average agro-ecosystem typically far outweighs the number of harmful insect species. For example, in a study carried out in a single agro-ecosystem in rice fields in Indonesia, Settle *et al.* (1996) recorded 500 beneficial insect species and 130 pest species. Another 150 insect species were deemed "neutral" since they do no attack rice, although they served a very important role in the survival of predators when rice was lacking. Beetles have also been used to control water hyacinth invasions. Snout beetles (*Neochetina* spp.), imported from Australia, successfully controlled the water hyacinth in Lake Victoria (Wilson *et al.*, 2007).

BOX 2.2
Common insect products and services

Cochineal (carmine dye): scale insects
Honey: bees
Shellac (polish): various Hemiptera
Pollination: various insects
Propolis (natural medicine): bees
Resilin protein (for artery repair): human flea
Royal jelly (beauty products): bees
Silk: silkworms
Termite hills (architectural models): termites
Venom (treatment for inflammatory diseases): bees
Beeswax (cosmetics and candles): bees

2.1.2 Beneficial roles of insects for humans

Besides serving as sources of food, insects provide humans with a variety of other **valuable products** (Box 2.2). Honey and silk are the most commonly known insect products. Bees deliver about 1.2 million tonnes of commercial honey per year (FAO, 2009b), while silkworms produce more than 90 000 tonnes of silk (Yong-woo, 1999). Carmine, a red dye produced by scale insects (order Hemiptera), is used to colour foods, textiles and pharmaceuticals. Resilin, a rubber-like protein that enables insects to jump, has been used in medicine to repair arteries because of its elastic properties (Elvin *et al.*, 2005). Other medical applications include maggot therapy and the use of bee products – such as honey, propolis, royal jelly and venom – in treating traumatic and infected wounds and burns (van Huis, 2003a).

Insects have also inspired **technology and engineering** methods. The silk proteins of arthropods (e.g. spiders) are strong and elastic and have been used as biomaterials (Lewis, 1992). The unique structure of silk, its biocompatibility with living systems, its function as a tool for new materials engineering and its thermal stability are only a few of the features that make it a promising material for many clinical functions (Vepari and Kaplan, 2007). For example, researchers inserted a spider's dragline silk gene into goat DNA in such a way that the goats would make the silk protein in their milk. This "silk milk" could then be used to manufacture a weblike material. Chitosan, a material derived from chitin that makes up the exoskeleton of insects, has also been considered as a potential intelligent and biodegradable biobased polymer for food packaging. Such natural packaging using the "skin" of insects can acclimatize the internal environment, protecting the product from food spoilers and micro-organisms. In particular, chitosan can store antioxidants and exhibits antimicrobial activity against bacteria, moulds and yeasts (Cutter, 2006; Portes *et al.*, 2009). However, the chitosan polymer is sensitive to moisture and could therefore be impractical in its 100 percent natural form (Cutter, 2006). Termite hills and their complicated network of tunnels and ventilation systems serve as useful models for constructing buildings in which air quality, temperature and humidity can be regulated efficiently (Turner and Soar, 2008). Drawing on nature – or rather imitating it – to solve human problems is called **biomimicry**.

The branch of entomology – or the scientific study of insects – that explores the influence of insects on culture (e.g. language, literature, art and religion) is known as **cultural entomology** (Box 2.3) (Hogue, 1987). Contributions from this field have helped highlight the distinct role that insects have assumed in literature (in particular children's books), movies and visual art, as well as their place as collection items, ornaments and more generally as inspiration for creative expression.

BOX 2.3

Examples of cultural entomology[2]

1. Ornamental insects

Insects are fascinating creatures that can easily be prepared and stored for extended periods. Given that some species and genera, mainly beetles and butterflies, are often large and extremely colourful, it is unsurprising that insects have become collectors' items. Species of commercial interest mostly belong to a few butterfly and beetle families. While the majority of insects are still collected in the wild, a number of butterfly farms rear and sell the pupae of common butterfly species. The rearing of insects by hobbyists is a recent trend (a few decades at best). The market for dead specimens and the number of insect collectors, however, seem to be in decline.

The public sector (e.g. zoological gardens and butterfly parks) is interested mainly in large and eye-catching butterfly species. These species are farmed or ranched in butterfly farms in tropical countries and the pupae shipped internationally.[3] Prices per pupae are low, ranging from a few cents up to a few United States dollars per pupa. These farms employ labour and provide income for local people. In Papua New Guinea, host to the fantastic-looking genus Ornithoptera, commonly known as bird wings, the government took an active approach in promoting butterfly ranching as a source of income for local farmers.

Beetles and other insects are not commonly reared in their countries of origin, but rather by private and professional insect growers around the world. Japan and Taiwan have strong beetle-farming communities (mainly for Lucanidae, Cetoniidae and Dynastidae), with the industrial-scale production of rearing materials, several insect shops

[2] The ornamental insects part of this box was contributed by Benjamin Harink.

[3] For a list of butterfly farms, please refer to the International Association of Butterfly Exhibitors.

Continues

Box 2.3 continued

in larger cities, and a number of magazines devoted to the topic of beetle-growing. There is however, good potential for rearing these insects in their countries of origin, for example by providing suitable sites for growing them.

The main challenges for the ornamental insect sector are legal. With declining forests and several species becoming extinct daily, more and more species are banned from trade. The second problem concerns public relations: insect collectors do not have a good reputation. They are often criticized for collecting live animals and keeping exotic species encaged as pets. Furthermore, increasing fears of invasive species complicates the transport of living animals across borders. In such a political climate, it might be better to support or create in-country agencies, such as the Insect Farming and Trading Agency of Papua New Guinea, to regulate the insect trade and ensure certain levels of income for insect growers and collectors, which would also serve as an incentive to protect natural forests.

Once the conditions have been determined, it is amazing how easily many ornamental species can be reared. Ornamental insect farmers can provide ideas for good species for human consumption to add to the existing range, such as the larvae of Dynastidae, which have the potential to grow up to 200 grams (g) each. Aquatic beetles can be consumed in both immature and adult stages (Ramos Elorduy, Pino and Martinez, 2008). Some Cetoniidae species have high reproduction and growth rates and might therefore have potential as human food, particularly as their larvae contain a lot of protein. This protein is packed in a soft skin, reducing the quantity of chitin that would need to be disposed of. The biggest advantage with these ornamental species, however, is that they live off compost and decomposing plant matter, like that produced by mushroom farms. Compost material left behind after the mushroom harvest would be an ideal substrate to inoculate with larvae of beetles for protein production. Moreover, there is no competition for food resources that could be used by humans. In addition, the excrement produced by the larvae makes good fertilizer and helps soil retain moisture for plants. Close cooperation between hobbyist insect growers and those interested in using insects as food is thus desirable, in the hope that new and more suited species than those currently in use might be discovered.

2. Singing crickets

Keeping crickets and bush crickets as pets is a centuries-old tradition in Asian cultures and even in some Western societies. Insects were first cited in an epigram dating to 600 BCE found in Ancient Greece, which referred to a young girl and her dying pet cricket. Many more poems have been written since, particularly on the sounds or songs of crickets (Weidner, 1952).

In China, singing crickets became domestic pets over 2 000 years ago. During the Tang Dynasty (618–906 CE), people kept crickets in cages to listen to their song:

> Whenever the autumn arrives, the ladies of the palace catch crickets and keep them in small golden cages, which are placed near their pillows so as to hear their songs during the night. This custom was also mirrored by common people.
> (Kai Yuan Tian Boa Yi Shi, *Affairs of the Period of Tian Bao*, 742–759 CE)

3. Cricket fighting

Cricket fighting flourished as a popular sport in China under the Song Dynasty (960–1278 CE). The practice was forbidden during the Qing Dynasty (1644–1911 CE), and cricket fighting became clandestine. Today, cricket fighting is again widespread, although mainly in large cities like Shanghai, Beijing, Tianjin, Guangzhou and Hong Kong, where cricket-fighting clubs and societies thrive. With the migration of Chinese people to other parts of the world, cricket fighting can now be found in places like New York and Philadelphia (Xing-Bao and Kai-Ling, 1994).

Continues

Box 2.3 continued

However, there is a negative side to cricket fighting, as overcollection has caused significant problems. In Shanghai alone there are 300 000–400 000 cricket enthusiasts, and about 90 percent of them are interested in betting on cricket fights. Crickets have become less abundant in areas around large cities in China. Collectors have even reportedly caused damage to suburban vegetable plots while searching for crickets. For more information see also Ryan (1996) and Costa-Neto (2003).

Source: Jin, 1998.

2.2 ENTOMOPHAGY AROUND THE WORLD

2.2.1 Number of edible insect species identified

Providing definitive figures on the number of edible insect species worldwide is difficult for several reasons. First, a layperson is unlikely to describe an insect by its Linnaean nomenclature, making official estimates difficult. Matters are complicated by the use in many cultures of more than one vernacular name – also called ethnospecies – for the same insect species. By using only Latin names and correcting for synonyms, Yde Jongema of WUR conducted a worldwide inventory using the literature, including from Western countries and temperate regions, and listed 1 900 edible insect species worldwide as of April 2012. Lower estimates do exist. DeFoliart (1997) counted a "low" 1 000 species, while Ramos Elorduy (2005) noted "at least" 1 681 species. Regional and national estimates have also been made: van Huis (2005) identified 250 edible species in Africa; Ramos Elorduy *et al.* (2008) listed 549 species in Mexico (although Cerritos, 2009, reported a mere 177 species in the country); in China, Chen *et al.* (2009) documented 170 species; Young-Aree and Viwatpanich (2005) reported 164 species in the Lao People's Democratic Republic, Myanmar, Thailand and Viet Nam; and Paoletti and Dufour (2005) estimated that 428 species were consumed as food in the Amazon (Figure 2.1).

FIGURE 2.1
Recorded number of edible insect species, by country

Source: Centre for Geo Information, Wageningen University, based on data compiled by Jongema, 2012.

2.2.2 Major groups of edible insects

Globally, the most common insects consumed[4] are beetles (Coleoptera) (31 percent) (see Figure 2.2). This is not surprising given that the group contains about 40 percent of all known insect species. The consumption of caterpillars (Lepidoptera), especially popular in sub-Saharan Africa (Box 2.4), is estimated at 18 percent. Bees, wasps and ants (Hymenoptera) come in third at 14 percent (these insects are especially common in Latin America). Following these are grasshoppers, locusts and crickets (Orthoptera) (13 percent); cicadas, leafhoppers, planthoppers, scale insects and true bugs (Hemiptera) (10 percent); termites (Isoptera) (3 percent); dragonflies (Odonata) (3 percent); flies (Diptera) (2 percent); and other orders (5 percent). Lepidoptera are consumed almost entirely as caterpillars and Hymenoptera are consumed mostly in their larval or pupal stages. Both adults and larvae of the Coleoptera order are eaten, while the Orthoptera, Homoptera, Isoptera and Hemiptera orders are mostly eaten in the mature stage (Cerritos, 2009).

FIGURE 2.2
Number of insect species, by order, consumed worldwide

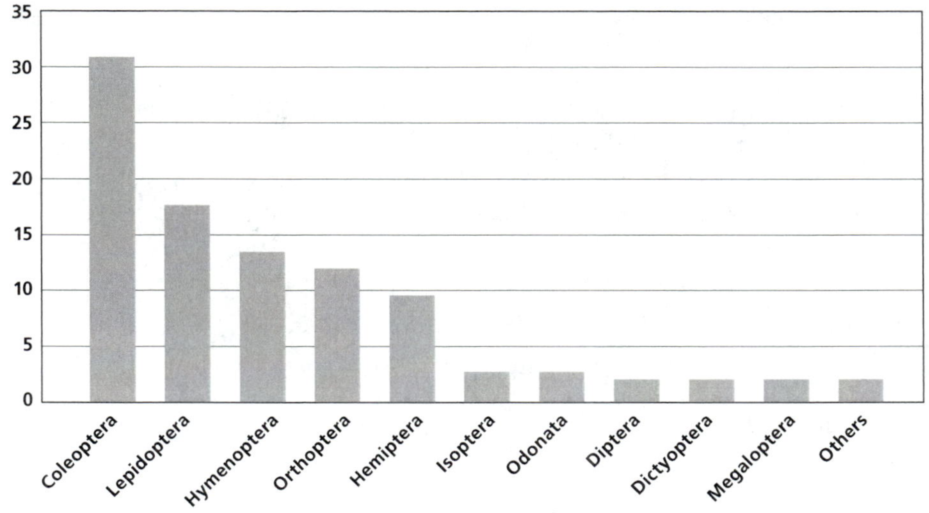

Note: total number = 1 909.
Source: Jongema, 2012.

BOX 2.4
Example of national insect diversity: species eaten in the Central African Republic

There is considerable variation in the most consumed insect order by continent, country and community. For example, an estimated 96 insect species are eaten in the Central African Republic. Orthoptera (locusts and grasshoppers) is the most consumed class (40 percent), followed by Lepidoptera (caterpillars) (36 percent), Isoptera (termites) (10 percent), Coleoptera (beetles) (6 percent) and others such as cicadas and crickets (8 percent).

Source: Roulon-Doko, 1998.

[4] This should not be confused with the frequency at which insects are eaten in a certain group.

Coleoptera (beetles)

There are many kinds of edible beetles, including aquatic beetles, wood-boring larvae, and dung beetles (larvae and adults). Ramos Elorduy, Pino and Martinez-Camacho (2009) listed 78 edible aquatic beetle species, mainly belonging to the Dytiscidae, Gyrinidae and Hydrophilidae families. Typically, only the larvae of these species are eaten. The most popular edible beetle in the tropics, by far, is the palm weevil, *Rynchophorus*, a significant palm pest distributed throughout Africa, southern Asia and South America. The palm weevil *R. phoenicis* is found in tropical and equatorial Africa (see Box 2.5 on the use of sound in harvesting), *R. ferrugineus* in Asia (Indonesia, Japan, Malaysia, Papua New Guinea, the Philippines and Thailand) and *R. palmarum* in the tropical Americas (Central America and West Indies, Mexico and South America).

In the Netherlands, the larvae of mealworm species from the Tenebrionidae family, such as the yellow mealworm (*Tenebrio molitor*), the lesser mealworm (*Alphitobius diaperinus*) and the superworm (*Zophobas morio*), are reared as feed for reptile, fish and avian pets. They are also considered particularly fit for human consumption and are offered as human food in specialized shops.

BOX 2.5

Use of sound in harvesting larvae

In Cameroon, women are generally involved in harvesting beetle larvae. They detect the larvae in palm trees by placing their ears against the tree and listening to the sound made by nibbling larvae. This method is commonly used to determine the optimal timing for harvesting the most sought-after instar (the developmental stage of an insect or larvae) of *Rhynchophorus* larvae. In the Democratic Republic of the Congo, the same method is used to harvest edible larvae of weevil, longhorn and scarab beetles, which occur in standing or rotting *Elaeis, Raphia, Chamaerops* and *Cocos nucifera* palm trees (Ghesquière, 1947).

Source: van Huis, 2003b.

Lepidoptera (butterflies and moths)

Butterflies and moths are typically consumed during their larval stages (i.e. as caterpillars), but adult butterflies and moths are also eaten. Indigenous Australians have been reported to eat moths of the cutworm *Agrotis infusa* (the Bogong moth) (Flood, 1980) and, in the Lao People's Democratic Republic, people have been observed eating hawkmoths (*Daphnis* spp. and *Theretra* spp.) after removing the wings and legs (J. Van Itterbeeck, personal communication, 2012). Nevertheless, the practice is limited.

The mopane caterpillar (*Imbrasia belina*) is arguably the most popular and economically important caterpillar consumed. Endemic to the mopane woodlands in Angola, Botswana, Mozambique, Namibia, South Africa, Zambia and Zimbabwe, the caterpillar's habitat extends over about 384 000 km^2 of forest (FAO, 2003). An estimated 9.5 billion mopane caterpillars are harvested annually in southern Africa, a practice worth US$85 million (Ghazoul, 2006). Other caterpillars are also consumed, but to a lesser extent. Malaisse (1997) identified 38 different species of caterpillar across the Democratic Republic of the Congo, Zambia and Zimbabwe. Latham (2003) documented 23 edible species in the Bas-Congo, a western province of the Democratic Republic of the Congo.

Caterpillar harvesting is not exclusive to Africa. In Asia, the bamboo caterpillar (*Omphisa fuscidentalis*), also known as the bamboo borer or bamboo worm, is a popular food that is being promoted by the Thai Department of Forestry of the Ministry of

Agriculture and Cooperatives as an increasingly viable source of income (Yhoung-Aree and Viwatpanich, 2005). In the Chiapas region in Mexico, locals are believed to consume up to 27 caterpillar species (Box 2.6).

BOX 2.6

Maguey worms

Red maguey worms – larvae of the moth *Comadia redtenbacheri* – and white maguey worms – larvae of the butterfly *Aegiale hesperiaris* – are found throughout central Mexico on the leaves of *Agave salmiana*. When fully mature, the highly nutritious caterpillars are considered a delicacy by Mexican farmers. They are generally eaten deep fried or braised, seasoned with a spicy sauce and served in a tortilla. Along with the larvae of the agave weevil (*Scyphophorus acupunctatus*), red maguey worms are one of the types of *gusano* (caterpillar) found in bottles of mezcal liquor (a distilled alcoholic beverage made from the maguey plant, *Agave americana*) in the Mexican state of Oaxaca. The *gusanos* are so popular that mezcal producers send security guards into agave fields during the rainy season to stop poachers.

Source: Ramos Elorduy *et al.*, 2007.

Hymenoptera (wasps, bees and ants)

Ants are highly sought-after delicacies in many parts of the world (Rastogi, 2011; Del Toro, Ribbons and Pelini, 2012). They also render important ecological services, including nutrient cycling, and serve as predators of pests in orchards, although negative effects are also reported (Del Toro, Ribbons and Pelini, 2012). The weaver ant (*Oecophylla* spp.) is used as a biological control agent in various crops, such as mangoes (Van Mele, 2008), and the larvae and pupae of the reproductive form (queen brood), also called ant eggs, constitute a popular food in Asia (see section 4.5.1). In Thailand they are sold in cans. Shen, Li and Ren (2006) reported that the black weaver ant (*Polymachis dives*) is widely distributed in subtropical southeast China, Bangladesh, India, Malaysia and Sri Lanka. It is used as a nutritional ingredient and processed into various tonics or health foods available on the Chinese market. The State Food and Drug Administration and State Health Ministry of China have approved more than 30 ant-containing health products since 1996.

In Japan, the larvae of yellow jacket wasps (*Vespula* and *Dolichovespula* spp.), locally known as hebo, are commonly consumed. During the annual Hebo Festival, food products made from the larvae of the wasps are popular delicacies (Nonaka, Sivilay and Boulidam, 2008), so much so that the local supply is insufficient and imports from Australia and Viet Nam are necessary to keep up with demand (K. Shono, personal communication, 2012). Box 2.7 provides background information on bees.

An inventory compiled by Ramos Elorduy and Pino (2002) in Chiapas, Mexico, suggested that most (67) insect species eaten in the state belong to the Hymenoptera order, and two leafcutter ant species (*Atta mexicana* and *A. cephalotus*) are becoming increasingly commercialized there. Further south, Amerindians have also been documented eating ants of the *Atta* genus (Dufour, 1987). Colonies of *Atta* species can have more than 1 million workers, and some can have up to 7 million. Their effect on vegetation in the Neotropics is said to be comparable with that of large grazing mammals on the African savannah. Therefore, a leafcutter colony can be considered competitive with a cow (Hölldobler and Wilson, 2010).

BOX 2.7
Beekeeping around the world

The contribution of bees to nature and agriculture is well documented (Bradbear, 2009), but their enormous potential to act as a direct source of food for humans is less understood (Chen *et al.*, 1998). A limited number of studies have shown that bee brood (eggs, larvae and pupae) and adults of a number of bee families are edible, including Bombycidae, Meliponidae and Apidae (Banjo, Lawal and Songonuga, 2006; Ramos Elorduy, 2006). An extensive nutritional analysis conducted by Finke (2005) showed that bee brood (presumably of *Apis mellifera*) is an excellent source of energy, amino acids, essential minerals and B-vitamins.

Nest-building insects, such as honeybees, lend themselves easily to semi-cultivation: bees can be attracted to nest at certain spots, and their hives can be relocated nearer to home, for example. These techniques have been applied widely worldwide over a long period (DeFoliart, 1995); in Central America, they date back to the Mayan civilizations (Villanueva, Roubik and Colli-Ucan, 2005). Coletto-Silva (2005) documents an ingenious method for collecting stingless bee (*Melipona* spp.) colonies to start meliponaries, whereby the host tree is not destroyed: the tree is opened up, the colony collected and the tree closed again with natural resins.

More facts on bees:

- Along with wasps, honeybees (*Apis mellifera*) are the most important food insects in northern Thailand. Bee brood features commonly in local diets and is in high demand in markets; therefore, it is often expensive (Chen *et al.*, 1998).
- In Malawi, beekeeping is more than three times as profitable as growing maize, a staple crop (Munthali and Mughogho, 1992).
- In Australia, the hive (referred to as honeybag or sugarbag) of native stingless bees (*Trigona* spp.) is a popular source of sugar for Aborigines (Cherry, 1991; O'Dea *et al.*, 1991).

Orthoptera (locusts, grasshoppers and crickets)

About 80 grasshopper species are consumed worldwide, and the large majority of grasshopper species are edible. Locusts may occur in swarms, which makes them particulary easy to harvest. In Africa, the desert locust, the migratory locust, the red locust and the brown locust are eaten. However, due to their status as agricultural pests they may be sprayed with insecticides in governmental control programmes or by farmers. For example, relatively high concentrations of residues of organophosphorus pesticides were detected in locusts collected for food in Kuwait (Saeed, Dagga and Saraf, 1993).

Grasshoppers and locusts are generally collected in the morning when the temperature is cooler (and the insects, being cold-blooded, are relatively immobile). In Madagascar, there is a common saying: "*Comment pourriez-vous attraper les sauterelles pondeuses et faire la grasse matinée en même temps?*" ("one needs to waken early in the morning to catch grasshoppers"). In Oaxaca, the harvest of chapulines (edible grasshoppers of the genus *Sphenarium*) only takes place very early in the morning (04:00–05:00 hours) (Cerritos and Cano-Santana, 2008) because chapulines are too active and difficult to catch during the hotter part of the day (Cohen, Sanchez and Montiel-ishinoet, 2009).

In the West African nation of Niger, it is not uncommon to find grasshoppers for sale in local markets or sold as snacks on roadsides. Remarkably, researchers found that grasshoppers collected in millet fields fetched a higher price in local markets than millet (van Huis, 2003b).

The chapuline is probably the best-known edible grasshopper in Latin America. This small grasshopper has been a part of local diets for centuries and is still eaten in several parts of Mexico. The valleys of Oaxaca state are especially famous for the consumption of chapulines. Cleaned and toasted in a little oil with garlic, lemon and salt for flavour, they are a common food ingredient among not only indigenous communities but also the urban population in Oaxaca city (Cohen *et al.*, 2009). Chapulines are brachypterous, which means they have reduced, non-functional wings. *Sphenarium purpurascens* is a pest of alfalfa but also one of the most important edible insects in Mexico. Harvesters use conical nets (about 80 cm in diameter and 90 cm deep) without handles to lightly beat the alfalfa plants, allowing each local family to obtain about 50–70 kilograms (kg) of grasshoppers weekly (Cerritos and Cano-Santana, 2008). Chapulines play a significant role in local small-scale markets as well as in restaurants and export markets. Despite the nutritional and cultural value of chapulines, recent studies have shown that the grasshoppers can contain high and sometimes dangerous amounts of lead (Cohen, Sanchez and Montiel-ishinoet, 2009).

In Asia, the crickets *Gryllus bimaculatus*, *Teleogryllus occipitalis* and *T. mitratus* are harvested in the wild and commonly consumed as food. The house cricket (*Acheta domesticus*) is also reared and commonly eaten, particularly in Thailand, and is preferred over other species because of its soft body. In a study carried out in Thailand in 2002, 53 of 76 provinces had cricket farms (Yhoung-Aree and Viwatpanich, 2005). As of 2012, there were about 20 000 cricket farmers in Thailand. Additionally, the short-tail cricket (*Brachytrupes portentosus*), which has a large body and large head, is also quite popular for eating. However, this species cannot be farmed and therefore is only collected in the wild (Y. Hanboonsong, personal communication, 2012).

Despite the extensive practice of farming insects, only two species of edible cricket (*Gryllus bimaculatus* and *Acheta domesticus*) are farmed economically. Others, such as *Tarbinskiellus portentosus*, cannot be farmed due to their long life cycles. However, there are signs of change in the Lao People's Democratic Republic and Cambodia: sellers are now saying that consumers prefer farmed crickets over those collected in the wild because they taste better (P. Durst, personal communication, 2012).

Homoptera (cicadas, leafhoppers, planthoppers and scale insects), a suborder of the Hemiptera

In Malawi, several cicada species (*Ioba*, *Platypleura* and *Pycna*) are highly esteemed as food. Cicadas can be found on the trunks of trees and collected using long reeds (*Phragmites mauritianus*) or grasses (*Pennisetum purpureum*) with a glue-like residue on them, such as latex from the *Ficus natalensis* tree. The latex adheres to the cicadas' wings, which are removed before consumption. Some Homoptera yield products commonly eaten by humans, such as carmine dye (a bright red pigment also called E120) derived from the cactus cochineal bug (*Dactylopius coccus*) often used in food products. Humans also consume lerp, a crystallized, sugary secretion produced by the larvae of psyllid insects as a protective cover. In South Africa, for example, the psyllid (*Arytaina mopane*) that feeds on the phloem sap of the mopane tree (*Colophospermum mopane*) is eaten. The largest number of lerp-building psyllids is found on *Eucalyptus* species in Australia. Australian Aborigines collect lerp as a sweet food source (Yen, 2005). See section 2.4.3 for more information on lerp.

Heteroptera (true bugs), suborder of Hemiptera

Pentatomid bugs are eaten widely throughout sub-Saharan Africa, particularly in southern Africa (see section 2.4.4). In the Republic of Sudan, the pentatomid *Agonoscelis versicolor*, a pest of rainfed sorghum that causes considerable damage, is eaten roasted. Oil is also derived from these insects and is used in preparing foods and for treating scab disease in camels (van Huis, 2003a).

Most pentatomids consumed as food, however, live in water. The famous Mexican caviar, ahuahutle, is composed of the eggs of at least seven species[5] of aquatic Hemiptera (the *Corixidae* and *Notonectidae* families); these insects have formed the backbone of aquatic farming, or aquaculture, in Mexico for centuries (Box 2.8). The semi-cultivation of these species is simple and inexpensive because it can be undertaken using traditional local practices (Parsons, 2010) (see Chapter 4). The insects fetch high prices, particularly during the *Semana Santa* (the week preceding Easter). The semi-cultivation of Hemiptera is under threat, however, as a result of heavy pollution and dried-up water bodies (Ramos Elorduy, 2006).

BOX 2.8

Ahuahutle, Mexican caviar

In *Historia de las cosas de la Nueva España*, Sahugan (1557) stated that at the court of Emperor Montezuma and the Aztec kings that preceded him prior to the tenth century, the ahuahutle were prepared especially during ceremonies dedicated to the god Xiuhtecutli. Native runners brought the ahuahutle into Tenochtitlan from Texcoco so that the Emperor could have them fresh for breakfast. Sahugan called them aoauhtli or ahuauhtli and reported that the common name used by the people was *aguaucle*, which meant "seeds of the water". He also reported that they were eggs deposited by flies on the surface of stagnant waters in infinite numbers and were sold in the marketplace of Texcoco and other neighbouring villages.

Source: Bachstez and Aragon, 1945.

Isoptera (termites)

The most commonly eaten termite species are the large *Macrotermes* species. The winged termites emerge after the first rains fall at the end of the dry season, from holes near termite nests. van Huis (2003b) observed that, in Africa, locals beat the ground around termite hills (simulating heavy rain) to provoke the termites to emerge.

Syntermes species are the largest termites eaten in the Amazon. They are gathered by introducing a palm leaf rib into the galleries of the nest; the soldiers biting it are then fished out (Paoletti *et al.*, 2003; Paoletti and Dufour, 2005). More information on termites is provided in section 2.3.3.

2.2.3 Where and when are insects eaten?

The frequency of insect consumption around the world is poorly documented. The few examples found in literature are from Africa, Asia and Latin America.

Africa

Insects can be found in abundance throughout the African continent and when staples are scarce they become important sources of food. During the rainy season – when hunting game or fish can be problematic – insects play an important role in food security. Caterpillars are especially popular during the rainy season, although their availability can vary even within the same country depending on climatic conditions (Vantomme, Gohler and N'Deckere-Ziangba, 2004); Table 2.1 shows the seasonal abundance of caterpillars in Central Africa.

The seasonal availability and correlated consumption of insects is well documented by Takeda and Sato (1993). A study carried out in tropical rainforest in the Democratic Republic of the Congo shows the remarkable resourcefulness of the Ngandu people, who obtain nourishment from what is seasonally available: cultivated and wild-gathered

[5] *Corisella mercenaria* (Say), *C. texcocana* (Jacz), *Krizousacorixa femorata* (Guér), *K. azteca* (Jacz), *Graptocorixa abdominalis* (Say), *G. bimaculata* (Guér) (Hemiptera-Corixidae) and *Notonecta* spp. (Hemiptera-Notonectidae).

plants, mushrooms, mammals, birds, fish, reptiles and insects. An earlier study carried out in the same country found that the availability of caterpillars is strongly correlated with declines of fish and game (Pagezy, 1975) (see Figure 2.3).

FIGURE 2.3

Monthly rainfall (top) and monthly occurrence of meals of fish, caterpillars and game in 15 consecutive months in the Lake Tumba region, Democratic Republic of the Congo

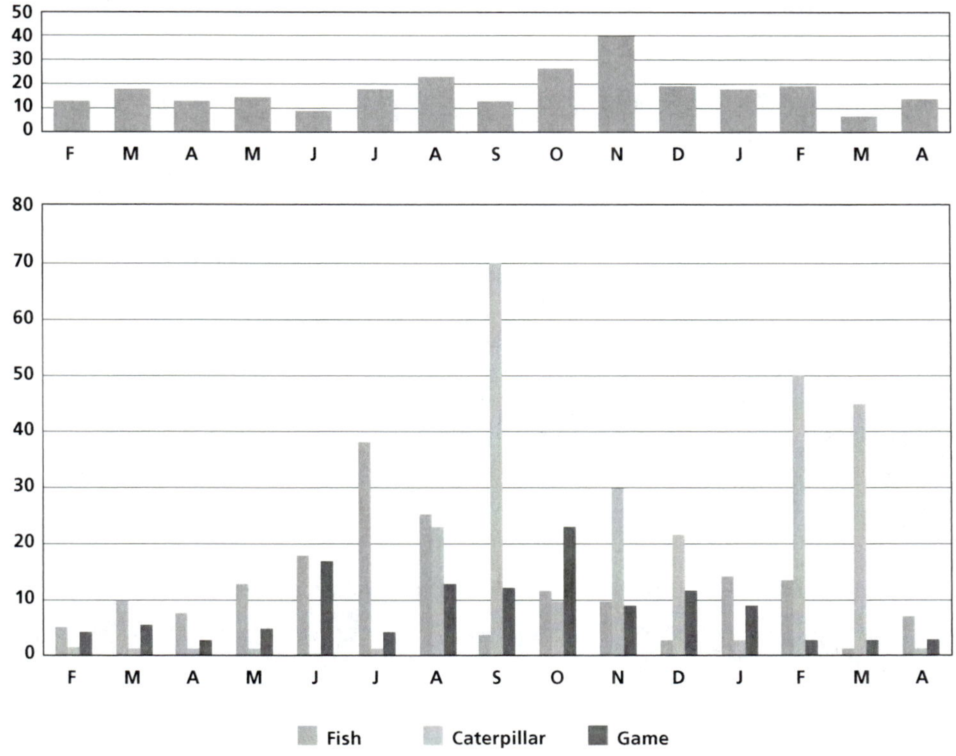

Source: Pagezy, 1975.

Markets in Kinshasa, the capital of the Democratic Republic of the Congo, boast an abundant year-round supply of caterpillars, and the average household in Kinshasa eats approximately 300 g of caterpillars per week. It has been estimated that 96 tonnes of caterpillars are consumed in the city annually (Kitsa, 1989). Consumption of the mopane caterpillar by far exceeds that of other caterpillars: 70 percent of Kinshasa's 8 million inhabitants are estimated to eat the caterpillars, for both their nutritional value and their taste (Vantomme, Gohler and N'Deckere-Ziangba., 2004).

Caterpillars also provide an important source of protein during the rainy season (July to October) in the Central African Republic (Bahuchet, 1975; Bahuchet and Garine, 1990), particularly for pygmies. In the rainy season, average consumption is estimated at 42 freshly harvested caterpillars per person per day. Consumption in the remainder of the year is much lower, although the insects are available year-round, either dried or smoked (see Figure 2.3). The indigenous Gbaya have been documented to consume 96 different insect species; this amounts to 15 percent of their protein intake (Roulon-Doko, 1998).

In some places, the consumption of insects is correlated with the availability of staples. In Madagascar, the consumption of rice declines at the end of the dry season and the consumption of caterpillars rises (Decary, 1937). Locals harvest caterpillars from forest trees at the end of the dry season as leaves develop just before the rain. The caterpillars can also be dried and stored for use in times of food shortage. In southern Africa, emperor moth caterpillars (Saturniidae) are widely consumed during food-deficient periods of the year.

TABLE 2.1
Abundance of caterpillars in Central Africa

Country	Province	Jan	Feb	Mar	Apr	May	June	July	Aug	Sep	Oct	Nov	Dec
CAR							■	■	■	■			
Cameroon								■	■	■	■		
DR Congo	East Kasaï						■	■					
	West Kasaï							■	■	■			
	Bandundu									■	■	■	■
	Kinshasa									■	■	■	■
Rep. du Congo	Sangha								■	■			
	Likoula								■	■			
	Brazzaville										■	■	■
	Pool	■											
	Plateaux	■											

Source: Roulon-Doko, 1998.

Asia

Between 150 and 200 species of edible insect are consumed in Southeast Asia. Red palm weevils (*Rhynchophorus ferrugineus*) from the Sago palm (*Metroxylon sagu*) are especially popular across the continent and are a highly prized delicacy in many regions (Johnson, 2010). Some insects are available year-round, including many aquatic species, while others are only available on a seasonal basis. Table 2.2 shows the annual availability of

TABLE 2.2
Availability of edible insects, Lao People's Democratic Republic, by month

Habitat	Common name (scientific name)	Jan	Feb	Mar	Apr	May	June	July	Aug	Sep	Oct	Nov	Dec
Aquatic	Water scorpion (*Laccotrephus* sp.) (Nepidae)	■	■	■	■	■	■	■	■	■	■	■	■
	Diving beetle (*Cybister* sp.) (Dytiscidae)	■	■	■	■	■	■	■	■	■	■	■	■
	Water scavenger (*Hydrophilus* sp.) (Hydrophilidae)	■	■	■	■	■	■	■	■	■	■	■	■
	Dragonfly larvae	■	■	■	■	■	■	■	■	■	■	■	■
	Giant water bug (*Lethocerus indicus*) (Belostomatidae)	■	■	■	■	■	■	■	■	■	■	■	■
Ground	Cricket (*Tarbinskiellus portentosus*) (=*Brachytrupes achatinus*) (Gryllidae)										■	■	■
	Scale insect (*Drosicha* sp.) [Monophlebidae = Margarodidae]	■	■	■									
	Dung beetles (Scarabaeinae)	■	■	■	■	■						■	■
Tree/ Bush/ Shrub	Cicada (Cicadidae)			■	■	■							
	Weaver ant (*Oecophylla smaragdina*) (Formicidae)			■	■	■							
	Stink bug (*Tessaratoma quadrata*) (Pentatomidae)				■	■	■						
	Scarabid beetle (*Holotrichia* sp.) (Scarabaeidae)				■	■	■						
	Grasshoppers (Orthoptera)					■	■	■	■	■			
	Bamboo caterpillar (*Omphisa fuscidentalis*) (Pyralidae)							■	■	■	■		

Source: Nonaka, 2010.

selected insect species in the Lao People's Democratic Republic. There, and in Myanmar, Thailand and Viet Nam, various insect species are collected throughout the year from a range of habitats; in this way, people can obtain a steady supply of edible insects (Yhoung-Aree and Viwatpanich, 2005) (see Table 2.3).

TABLE 2.3
Availability of edible insects in Thailand, by month

Month	Insect
January	Grasshopper, tortoise beetle, skipper
February	Adult red ant, dung beetle, scarab beetle, stink bug
March	Cicada, termites, dung beetles
April	Dung beetle, grasshopper
May	Ground cricket
June	Giant water bug, wood-boring beetle, predaceous diving beetle
July	Back swimmer, crawling water beetle, damselfly, spider
August	Bee hornet, wasp, beetle
September	Rhinoceros beetle, spider
October	Cricket
November	Long-horned beetle
December	Mole cricket, river swimmer, true water beetle, water scavenger beetle, water scorpion beetle

Source: Yhoung-Aree and Viwatpanich, 2005.

Present day entomophagy in many Asian countries is the result of migration patterns. For example, insects have long been an important part of diets in northeastern Thailand, but as a result of labour-migration to tourist areas in southern parts of the country, including Bangkok, the practice is now well established throughout the country (Yen, 2009); it is estimated that as many as 81 insect species are consumed in both rural and urban areas there. Additionally, over 50 insect species are consumed in South Asia (India, Pakistan and Sri Lanka), 39 species in Papua New Guinea and the Pacific Islands, and 150–200 species in Southeast Asia (Johnson, 2010).

Latin America

In Mexico, indigenous people possess a deep knowledge of the plant and animal species that traditionally make up their diets, including the life cycles of insects (Ramos Elorduy, 1997) (Box 2.9). Insects have been "calendarized" by species, meaning that they are believed to operate in harmony with natural phenomena such as plant life cycles, moon cycles, rainy seasons and thunder. It is widely known among indigenous people, for instance, that escamoles (larvae of the ants of the *Liometopum* genus) are ready to harvest when the jarilla plant (*Senecio salignus*) is flowering. In Oaxaca, Mexico, the harvest of chapulines begins with the onset of the rainy season and continues throughout this season. In the Amazon, insect gathering is also a seasonal affair. Maku Indians, an indigenous group of hunter-gatherers living in the tropical forest of northwestern Amazonia in Brazil, gather insects during the rainy season (from July to September) when hunting fish and game is difficult (Milton, 1984). In the Colombian Amazon, the Nukak community harvests larvae of *Rhynchophorus* species during the rainy season (Politis, 1996).

In the Ecuadorian highlands, the *Platycoelia lutescens* beetle can be found in Quito's markets from late October to early November; they are collected during winter rains. The beetles are harvested when they emerge from the soil in meadows and grasslands and are relatively easy to collect. It is believed that vibrations caused by rain and the sound of thunder trigger their emergence (Smith and Paucar, 2000). Not all insects, however, are harvested during the rainy season. For example, the larvae of the South American palm weevil (*Rhynchophorus palmarum*) and bearded weevil (*Rhinostomus barbirostris*)

BOX 2.9

Wild food consumption by the Popoloca people of Los Reyes Metzontla Puebla, Mexico

Food availability

Wild foods provide important supplements to the diet of the Popoloca people, particularly when maize and bean reserves are lean (Table 2.4). Wild plants and insect species are primarily available during the rainy season, from April to October, and before the harvest of maize and beans. The Popoloca typically gather wild foods as they go to work in their agricultural plots. In May, for example, several cacti fruits, such as chende (*Polaskia chende*), chichipe (*Polaskia chichipe*), xoconostle (*Stenocereus stellatus*), pitaya (*Stenocereus pruinosus*) and nopal de monte (*Opuntia depressa*), are harvested, along

Continued next page

TABLE 2.4

Available insect and insect products for the Popoloca people of Los Reyes Metzontla Puebla, Mexico

Month	Insect and insect products	Approximate amount of consumption
Year-round	Cazahaute worm	From ¼ to ½ litre two to three times a year per family
	Wasp nest (five species)	From 1 to 4 nests a year per family
	Apis mellifera honey	No data available
January	Year-round insects	-
February	*Comadia redtenbacheri*	Around 1 litre once or twice a year per family
March	Mormidea (Mormidea) *Notulata* and *Euschistus* sp.	From 1 to 2 litres once or twice a year per family
	Comadia redtenbacheri	Around 1 litre once or twice a year per family
	Plebeia mexica honey	Collected once a year during spring
April	Mormidea (mormidea) *Notulata* and *Euschistus* sp.	From 1 to 2 litres once or twice a year per family
	Thasus gigas	From ¼ to 2 litres one to three times a year per family
	Plebeia mexica honey	Collected once a year during spring
May	Mormidea (mormidea) *Notulata* and *Euschistus* sp.	From 1 to 2 litres once or twice a year per family
	Plebeia mexica honey	Collected once a year during spring
	Aegiale hesperiaris	Around 50 larvae each season per family
	Atta mexicana	From ¼ to 1 litre once a year per family
June	*Aegiale hesperiaris*	Around 50 larvae each season per family
	Pochocuile	Around one or two "medidas" (12 larvae) once a year per person
July	*Aegiale hesperiaris*	Around 50 larvae each season per family
	Pochocuile	Around one or two "medidas" (12 larvae) once a year per person
August	*Paradirphia fumosa*	From a couple of "medidas" (15 larvae) per person to 3 litres per family once a year
	Gusano del pirul	From ¼ to 1 litre two or three times a year per family
September	*Paradirphia fumosa*	From a couple of "medidas" (15 larvae) per person to 3 litres per family once a year
October	Year-round insects	-
November	Year-round insects	-
December	Year-round insects	-

Source: Acuña *et al.*, 2011.

Box 2.9 continued

FIGURE 2.4
Temporal availability of edible insects, wild plants and subsistence crops for the Popoloca people of Los Reyes Metzontla Puebla, Mexico

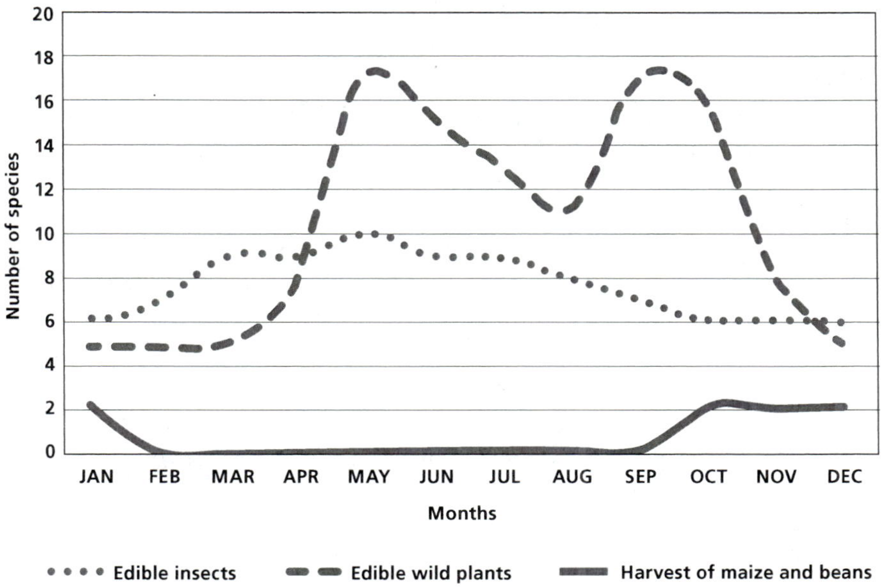

Source: Acuña *et al.*, 2011.

with insects such as chinches (stink bugs), conduchos (white agave maguey worm *Aegiale hesperiaris*) and chicatanas (the leafcutting ant *Atta mexicana*). Most of these insects (about 60 percent) are available from February to September, while about 40 percent are consumed year round (e.g. wasp nests).

Frequency and quantity
The frequency and quantity of insect consumption depend on three main factors: the consistency of climatic conditions, which can affect the quantity of insects harvested; personal choice; and chance encounters during other subsistence activities (e.g. farming) for species that are used opportunistically.

Source: Acuña *et al.*, 2011.

are collected by the Joïti people in northeastern Amazonia, Venezuela, from September to January at the end of the rainy season. Rain, in fact, keeps adult beetles away and increases the occurrence of fungal attack (Choo, Zent and Simpson, 2009).

2.3 EXAMPLES OF IMPORTANT INSECT SPECIES CONSUMED
This section describes some of the most consumed insect species but is by no means exhaustive.

2.3.1 Caterpillar
Caterpillars are among the world's most diverse groups of edible insects. They are not only valuable sources of protein and other micronutrients, they also make valuable contributions to livelihoods in many parts of the world. Among the most renowned are

the witchetty grubs[6] consumed in Australia (Meyer-Rochow, 2005) and the bamboo caterpillar (*Omphisa fuscidentalis*), which is popular in Thailand and the Lao People's Democratic Republic (Yhoung-Aree and Viwatpanich, 2005). The consumption of caterpillars is especially pervasive in sub-Saharan Africa, where 30 percent of all edible insect species are caterpillars (van Huis, 2003b). Malaisse (1997) listed 38 species of edible caterpillars on the basis of intensive studies in the region inhabited by the Bemba (Bantu-speaking people in the northeastern plateau of Zambia and neighbouring areas of the Democratic Republic of the Congo and Zimbabwe). In the Democratic Republic of the Congo, caterpillars make up 40 percent of the total animal protein consumed (Latham, 2003) (Box 2.10). The most popular and profitable caterpillar on the African continent is undoubtedly the mopane caterpillar, *Imbrasia* (=*Gonimbrasia*) *belina*.

Mopane caterpillar

The mopane woodlands are found in Botswana, Namibia, Zimbabwe and northern parts of South Africa. It is in this vast habitat that the mopane caterpillar thrives. Local knowledge of the insect's ecology and biology in some rural communities is extensive (Mbata, Chidumayo and Lwatula, 2002). Its distribution is largely correlated to that of its principal host, the mopane tree (*Colophospermum mopane*). The mopane caterpillar is bivoltine in most areas; that is, two generations are produced each year (the first between November and January, its major outbreak, and the second between March and May) (Stack *et al.*, 2003; Ghazoul, 2006).

Like many other edible insects, mopane caterpillars are not merely "famine foods", consumed in times of food shortage. Although the caterpillars are important sources of nutrition in lean times, they also form a regular part of the diet (Stack *et al.*, 2003).

BOX 2.10

Yansi sayings, Democratic Republic of the Congo

"Caterpillars and meat play the same role in the human body."
"As food, caterpillars are regulars in the village but meat is a stranger."

Source: Muyay, 1981.

Collecting, processing, trading and consuming the mopane caterpillar is an integral part of local cultures, but it is especially a livelihood strategy among marginalized groups (Illgner and Nel, 2000; Stack *et al.*, 2003). The caterpillars are collected by hand – primarily by women and children – and then degutted, boiled in salted water and sun-dried. Dried mopane caterpillars will last for several months and can be a valuable source of nutrition in times of stress. Harvesting and trading the caterpillars also provides important income for many rural families; this is often the prime incentive for harvesting (Stack *et al.*, 2003), and the income is comparable with and often higher than that generated by conventional agricultural crops (Munthali and Mughogho, 1992; Chidumayo and Mbata, 2002). The income generated by the mopane caterpillar harvest provides many families with funds to purchase household items such as clothing, school materials and basic utensils (Stack *et al.*, 2003; N'Gasse, 2004). Vast numbers of people partake in the mopane harvest: the nutritional and economic incentives are so high that many are willing to travel hundreds of kilometres across the mopane woodlands in search of the insects (Kozanayi and Frost, 2002).

[6] A term used in Australia for large, white, wood-eating larvae of the cossid moth (*Xyleutes* (=*Endoxyla*) *leucomochla*), a traditional Aboriginal delicacy.

The protein content of the mopane caterpillar is 48–61 percent and fat content is 16–20 percent, of which 40 percent is essential fatty acids. Mopane caterpillars are also a good source of calcium, zinc and iron (Glew *et al.*, 1999; Headings and Rahnema, 2002). See Chapter 6 for more information on nutrition.

2.3.2 Palm weevil

> *"Larvae assate in deliciis habentur"* [fried larvae are delicious] – Linneus on *Rynchophorous* spp. in the 1758 work *Systema Naturae.*

Larvae of the palm weevil (*Rynchophorous* spp.) are consumed in Asia (*R. ferrugineus*), Africa (*R. phoenicis*) and Latin America (*R. palmarum*). Their delicious flavour (Cerda *et al.*, 2001) is credited by some to their elevated fat content (Fasoranti and Ajiboye, 1993). In the tropics, the insects occur year-round where hosts are found. Often these hosts are trees under stress; that is, trees previously damaged by other insects, notably rhinoceros beetles (*Oryctes* spp.) or by the local traditional tapping for palm wine (Fasoranti and Ajiboye, 1993). Fallen palms can serve as breeding sites and support hundreds of larvae; for this reason, palms are often felled intentionally. Such a practice is common among Yanomamö (Chagnon, 1983) and Jöti Indians (Choo, Zent and Simpson, 2009) of the Amazon. Van Itterbeeck and van Huis (2012) noted that many indigenous people have excellent ecological knowledge of the palm weevil and can increase its availability and predictability through semi-cultivation practices. Experiments in Alto Orinoco villages have explored ways of making the production of palm weevils more sustainable than traditional palm cutting to improve the oviposition (egg-laying) of *R. palmarum* and other palm weevils (Cerda *et al.*, 2001).

Ecology

Palm weevils attack palm species, the most important of which are the coconut palm (*Cocos nucifera*), date palm (*Phoenix dactylifera*), sago palm (*Metroxylon sagu*), oil palm (*Elaeis guineensis*) and Raphia palm (*Raphia* spp.). For the African palm weevil, Fasoranti and Ajiboye (1993) noted that adult females lay a few hundred eggs on the new leaves of the plant or directly in the palm trunk. The weevil larvae burrow into the palm heart, causing its death. The total life cycle takes seven to ten weeks. A fully extended larva measures, on average, 10.5 cm in length and 5.5 cm in width and weighs 6.7 g. Extracting larvae from palms is labour-intensive and is often only carried out by young men (Fasoranti and Ajiboye, 1993).

BOX 2.11

Red palm weevil

The Red palm weevil, *Rhynchophorus ferrugineus*, is prevalent in most Asian countries and in the Middle East. It reached the Mediterranean in the 1980s as an invasive species, destroying over 13 000 date palms in Sicily by August 2009. The beetle has also spread along the Mediterranean coast and invaded mainland Italy and is killing trees as far north as Genoa. In Italy, its destruction is mainly limited to the ornamental plant *Phoenix canariensis*. The main control method in the country and throughout the Mediterranean is the systematic use of insecticides.

Source: Mormino, 2009.

Detecting the larvae

In the Democratic Republic of the Congo it is customary for women to detect the most opportune moment to harvest weevil larvae and longhorn and scarab beetles – which occur in standing or rotting *Elaeis*, *Raphia*, *Chamaerops* and *Cocos nucifera* palm trees (Ghesquière, 1947) – by putting their ears against the trees and listening to the sound made by the chewing and burrowing beetle larvae. This method is also used in Cameroon to harvest the palm weevil larvae (*Rhynchophorus phoenicis*) in its most appropriate instar (developmental stage) for consumption (van Huis, 2003b). The same practice has been documented in the Central African Republic (Roulon-Doko, 1998) and the Americas (Ghesquière, 1947; Wolcott, 1933). In Italy, forest inspectors have been known to use electronic listening devices to detect early infestations of the red palm weevil, given that when symptoms of the damage become apparent the palm tree will die (Box 2.11).

Consumption

Palm weevil larvae are typically collected, washed and fried for consumption (Fasoranti and Ajiboye, 1993). It is unusual to add oil because the larvae are high in fat and exude oil during the frying process. Common condiments include onion, pepper and salt. Barbecuing the larvae is also common practice.

In Nigeria, adults discourage children from eating palm weevil larvae. It is thought that this is done to prevent children from felling palm trees, which can increase breeding sites for the available stock of number of larvae to be harvested in the short term but would cause irrevocable long-term damage to host trees (Fasoranti and Ajiboye, 1993). Protecting palm trees is considered essential to communities, who depend on them for other key products, including palm oil, palm kernels and palm wine.

2.3.3 Termites

In the Western world, termites are generally synonymous with pests and are renowned for their capacity to devour wood. Damage from termites is said to cost over half a billion dollars per year in the United States of America alone. Yet termites are considered a delicacy in many parts of the world. They are consumed both as main and side dishes, or simply eaten as snack foods after they have been de-winged, fried and sun-dried (Kinyuru, Kenji and Njoroge, 2009).

Although they are often erroneously referred to as ants or white ants, termites occupy a different order to ants, viz. Isoptera. Edible termites, which typically belong to a family of macrotermites (Macrotermitinae), generally consist of the winged form that swarm from termite hills shortly after the first rains begin at the end of the dry season (often called the nuptial flight). These winged termites are the future queens and kings. They can be eaten, as can soldier termites. Termites are known to have large and elaborate nests; some species have nests as tall as 8 m, and a single nest may house as many as 1 million individuals consisting of workers, soldiers, a queen and a king. The global biomass of termite individuals is believed to exceed that of all human beings combined.

Termites cannot digest cellulose and lignin, so their digestive systems contain symbiotic protozoa and bacteria that digest the cellulose in wood. Termites live on the byproducts of this digestion and on the bodies of the symbionts themselves. For example, species of macrotermites use fungi in their nest, which aid in breaking down cellulose and lignin into a more nutritious source of food. The fungi are part of an extracorporeal digestive system that converts undigested woody material in plants into higher-quality oligosaccharides and more easily digestible complex sugars. In turn, termites "outsource" cellulose digestion. This digestion is responsible for about 4 percent of global GHG emissions from methane (Sanderson, 1996).

Queen and soldiers

Queen termites are considered particularly important delicacies, often reserved for special occasions (van Huis, 2003b). Their nutritional value is so high that in Uganda and Zambia they are fed to undernourished children. However, digging queens – which are capable of laying 2 000 eggs per day and measure up to 10 cm in diameter – is laborious, and their removal causes the death of entire colonies.

The consumption of soldiers of larger termite species has been documented in the Central African Republic, the Democratic Republic of the Congo, the Bolivarian Republic of Venezuela and Zimbabwe (Bequaert, 1921; Bergier, 1941; Owen, 1973; Chavanduka, 1976; Roulon-Doko, 1998; Paoletti *et al.*, 2003). They are often fried or pounded into cakes. Sometimes, for example in Uganda, only the heads are eaten (van Huis, 2003b). Termite soldiers can only be collected in small quantities, and generally collection is by women and children (Roulon-Doko, 1998). Unlike the winged forms, the soldiers can be gathered at any time of the year.

Collecting termites

Winged termites can be collected in a number of ways. In urban areas, they are trapped in receptacles with water near light sources, to which they are attracted. In rural areas, winged termites are typically caught at the termite mound itself. When they emerge – attracted by the light of a bundle of grass set on fire – they are swept into a hole dug for the purpose. In parts of the Democratic Republic of the Congo, people place baskets upside down over the holes so that the termites, which cling to the bottom of the baskets, fall into the holes when the baskets are shaken (Bergier, 1941). Instead of baskets, structures made of sticks or elephant grass covered with banana or maranta leaves or a blanket are also used to cover the holes (Bergier, 1941; Osmaston, 1951; Roulon-Doko, 1998). All escape routes are fenced off so that the termites are forced to emerge through a single opening on one side of the structure, to which the flying termites are attracted because of light from the sun, moon, torch or fire. A receptacle is placed near this opening to collect the termites (Harris, 1940; Bergier, 1941; Ogutu, 1986). Osmaston (1951) described how, in Uganda, intricate networks of clay pipes were assembled over the emergence holes, leading to a receptacle. It has also been reported that continuous beating and drumming on the ground (resembling rain) around termite hills triggers certain termite species to emerge (Owen, 1973; Ogutu, 1986; Roulon-Doko, 1998). Recently, Ayieko *et al.* (2011) combined modern technology with a indigenous practice for collecting termites (Box 2.12).

BOX 2.12

Merging traditional knowledge and new technologies for termite harvesting in Kenya

In Kenya, a study carried out in cooperation with Kenya Industrial Development found that constructing a simple light trap and receptor would facilitate the mass collection of *Macrotermes subhyalinus*, known in the Lake Victoria region as agoro, and lead to increased food security among those communities practising entomophagy.

The study proposed teaching communities in the region to construct traps using local and readily available materials in order to maximize collection and stressed the need to develop local familiarity with the various species of termites. For example, understanding the emergence pattern of the agoro termite – in short, identifying potential active mounds – would maximize collection. It would be equally important to take stock of changing environmental contexts. Merging modern science with indigenous practice shows promise, but further research is necessary to understand why there is so much variation in current yields, among other issues.

Source: Ayieko *et al.*, 2011.

Eating termites and nutritional value

Termites are rich in protein, fatty acids and other micronutrients. Fried or dried termites contain 32–38 percent proteins (Tihon, 1946; Santos Oliveira *et al.*, 1976; Nkouka, 1987). Essential fatty acids such as linoleic acid are particularly high in the African above-ground hill termite species, *Macrotermes bellicosus* (34 percent) and *M. subhyalinus* (43 percent) (Santos Oliveira *et al.*, 1976). In the Bolivarian Republic of Venezuela, soldiers of *Syntermes* species (e.g. *Syntermes aculeosus*) are renowned for their high nutritional value. The protein content of this species is a remarkable 64 percent; the genus is also rich in essential amino acids such as tryptophan, iron, calcium and other micronutrients.

Termites are generally consumed fried, sun-dried or smoked, although they are steamed in banana leaves in Uganda. To sun-dry or smoke termites they must first be killed by boiling or roasting for a few minutes (Silow, 1983). Sometimes they are crushed into powder form with a pestle and mortar and eaten with honey (Ogutu, 1986). The Azande people and pygmies in the Democratic Republic of the Congo fry meats in the fat residue of these termites (Bequaert, 1921; Bergier, 1941). The pygmies also use the oil to treat their body and hair. Termite oil is extracted by pressing dried termites in a tube (Costermans, 1955). In many East African towns and villages, sun-dried termites can be bought at local markets (Osmaston, 1951; Owen, 1973). Sun-dried termites can be ground into powder and mixed with other food ingredients (Pearce, 1997) by baking, boiling, steaming or processing them into crackers, muffins, sausages or meat loaves (Kinyuru, Kenji and Njoroge, 2009; Ayieko, Oriamo and Nyambuga, 2010). In Botswana, San women collect the winged termites *Hodotermes mossambicus* and roast them in hot ash and sand (Nonaka, 1996).

Termites as feed for pigs, poultry and fish

The use of termites as feed is documented in several countries. In Burkina Faso, termites are harvested using small calabashes, which are ingeniously filled with moist old dung, mango pits and other organic material and placed under the ground (van Huis, 1996). Three to four weeks later, the calabashes are unearthed and the contents – filled with termites – are fed to poultry. Such methods are particularly important at the end of the dry season when food is scarce (Iroko, 1982). Farina, Demey and Hardouin (1991) have shown how termites are fed to guinea fowls and chickens in villages in Togo using a technique comparable to that used in Burkina Faso. Swarming alates have been used to feed jungle fowl chicks in India and also ostriches in farms across Africa (Pearce, 1997).

Mushrooms from termite nests

In addition to termites, mushroom species found growing on termite nests are consumed regularly in many tropical countries. Wild mushrooms constitute important food supplements for local populations and also have a role in cultural traditions. In many parts of Africa, mushrooms are commonly found in markets, and are also stored for use during the cold dry season (Parent and Thoen, 1977). In Nigeria, Yoruba traditional doctors use a number of *Termitomyces* species (Lyophyllaceae) as medicines or charms. These mushrooms also have a place in mythical folklore (Oso, 1977).

Mushrooms belonging to the *Termitomyces* genus arise directly from the fungal combs in termite nests (Zoberi, 1973). The local names for these mushrooms are often derived from the local names for the termites. In Uganda, for example, the Nyoro tribe uses the term obunyanaka for mushrooms that grow on mounds of termite species known as enaka, while obunyantaike mushrooms grow on hills of the termite species entaike. Species of *Termitomyces* are large (up to 80 cm in diameter), with the exception of *T. microcarpus*, which is 0.5–2 cm in diameter (Parent and Thoen, 1977); this latter species is found across West Africa and in southern Africa (Skelton and Matanganyidze, 1981).

2.3.4 Stink bugs

Throughout Mexico (Ramos Elorduy and Pino, 2003), southern Africa and Southeast Asia, it is not uncommon to find people eating stink bug (Hemiptera: Pentatomidae) nymphs and adults (DeFoliart, 2002). In southern Africa, *Encosternum* (=*Natalicola*) *delegorguei* are considered delicacies. Stink bugs are consumed in Malawi, South Africa and Zimbabwe (Faure, 1944; van Huis, 2003b; Morris, 2004), while *Tessaratoma* species, *T. papillosa* (litchi stink bug), *T. javanica* (longan stink bug) and *T. quadrata* ("mien kieng", a local name in the Lao People's Democratic Republic) are widely sought-after in China, the Lao People's Democratic Republic and Thailand (Nonaka, 2007; Chen, Feng and Chen, 2009).

Ecology

Encosternum delegorguei, commonly known as the edible stink bug (thongolifha in the Venda language, xipembele in the Tsonga language and podile in the Northern Sotho language, southern Africa) are large, herbivorous, pale-green bugs with piercing and sucking mouthparts that they use to feed on plant juices (Triplehorn and Johnson, 2004). The name stink bug derives from the smell the insects release when they are disturbed (Aldrich, 1988). Stink bugs are collected from May to August, the period in which they occur in large numbers (Faure, 1944; Dzerefos, Witkowski and Toms, 2009). In Southeast Asia, aggregations of *Tessaratoma* species occur in a variety of trees during the dry season (peaking in March and April) (J. Van Itterbeeck, personal communication, 2012). Stink bugs also feed on crops and for this reason are considered agricultural pests (Panizzi, 1997).

Livelihoods

Inflated stink bugs or tessaratomids make an important contribution to rural diets in many parts of the world. In Zimbabwe, stink bugs are a valuable source of income for the Norumedzo community and are essential for buying household items and covering school fees (Makuku, 1993). Stink bugs are exported to neighbouring countries due to high demand, and collectors are known to travel up to 200 km to areas rich in stink bugs (Teffo, 2006).

Data on the nutritional value of stink bugs, however, are scarce, although some accounts do exist. According to Teffo (2006), *E. delegorguei* has a protein and fat content of 35.5 g per 100 g and 50.6 g per 100 g of edible weight, and consuming 100 g of *E. delegorguei* provides 2 599 kilojoules of energy. This species was also found to be high in minerals such as iron, potassium and phosphorus. In Southeast Asia, stink bugs of the *Tessaratoma* genus are especially esteemed. In the Vientiane Municipality, the Lao People's Democratic Republic, they are collected, consumed and sold *en masse* (J. Van Itterbeeck, personal communication, 2012).

Stink bugs are collected by hand throughout southern Africa, as well as in Southeast Asia. Collecting often results in yellow or orange staining from defensive secretions (Faure, 1944), which is why collectors tend to cover their hands with plastic bags (J. Van Itterbeeck, personal communication, 2012), and long sticks with nets attached to one end are also used. The insects are dislodged from trees by throwing small sticks or by shaking branches (J. Van Itterbeeck, personal communication, 2012). Collectors take special care to protect their eyes, as they believe secretions cause infections of the cornea and even blindness (Faure, 1944; Siripanthong *et al.*, 1991). It is easiest to collect the bugs when temperatures are cool – in the early morning, at sundown and especially after rain showers (Faure, 1944).

In both southern Africa and Southeast Asia, stink bugs are eaten both raw and cooked (Faure, 1944; J. Van Itterbeeck, personal communication, 2012). The heads of live or dead bugs are removed by squeezing (from back to front), which discards their "poison" (Faure, 1944; Toms and Thagwana, 2003). In the Lao People's Democratic Republic, only the scutellum (neckpiece) is removed after frying; it is said to be the source of its bitter taste (J. Van Itterbeeck, personal communication, 2012). Soaking the bugs in water, as well as immersing them in tepid water, also causes the insects to release their secretions; they can then be sun-dried for consumption (Toms and Thagwana, 2003). The water in which *Nezara robusta*, the green shield stink bug, leaves its secretions is used as a pesticide to protect houses and gardens from termites (Morris, 2004).

Ecological implications

Stink bugs face similar threats to many other highly sought-after edible insects. Because they have become a significant source of income and nutrition, over-harvesting and mismanagement of their habitats are increasing concerns. One reason for this is that many amateur collectors fell entire trees prior to harvest, with dangerous consequences for the sustainability of the practice (Faure, 1944; J. Van Itterbeeck, personal communication, 2012). In addition, overharvesting can and eventually will undermine stink bug populations, threatening subsequent copulation periods (beginning in mid October). Another issue has consequences for both the environment and food safety – stink bugs of the *Tessaratoma* genus are considered agricultural pests and may be subject to chemical treatment (e.g. the lychee stink bug, *Tessaratoma papillosa*, which is found on *Litchi chinensis*) (Menzel, 2002), which is a public health concern. The harvesting of stink bugs could protect crops and provide additional income and nutrition; eradication of the pest would likewise eradicate an important source of livelihoods, which should be avoided (Cerritos, 2009).

In some parts of the world, the benefits of insect gathering to lives and livelihoods are incentives for proper management. The Norumedzo community in Zimbabwe, for example, has designated stink bug habitats as community-protected areas. These forests are continually monitored and tree-felling is kept to a minimum (Makuku, 1993).

Some agricultural fields where stink bugs occur are subject to mechanical harvesting. In these cases, stink bugs are collected by hand to preserve the crop and earn income from the sale of the bugs. This method is becoming increasingly common in cultures where agricultural pests are also valuable sources of nutrition and income.

2.3.5 The edible grasshopper, *Ruspolia differens*

Development and collection

The edible grasshopper (*Ruspolia differens*), formally known as *Homorocoryphus nitidulus vicinus*, is a long-horned grasshopper of the Tettigoniidae family. It is a common food source in many parts of eastern and southern Africa. In the Lake Victoria region of East Africa, where the grasshoppers are known as nsenene, they form a major part of food culture (Kinyuru, Kenji and Muhoho, 2010). The Bahaya ethnic group in Tanzania's Bukoba district considers grasshoppers a delicacy. In Uganda, nsenene are traditionally collected by women and children.

Grasshopper eggs – which are laid in batches in the haulms of grasses – do not develop under dry conditions. Rainfall triggers development, which takes about four weeks (McCrae, 1982). Larvae and adults feed off grass anthers or grains such as rice, millet, sorghum and maize. Traditionally, grasshoppers are gathered during the day from these grasses (Mors, 1958).

"Okulinga ensenene" means that the Bahaya peoples (in Tanzania) step out of their huts in the early morning to look for nsenene in the fields. When they find them, they cry loudly to announce to the village where the nsenene have alighted – in banana groves or open fields, or on the hills. The young and old, especially women and children, go out to catch them. The grasshoppers may be gathered anywhere they fall, and owners of banana groves, for example, cannot expel as trespassers those who come to collect the ensenene. In collection times, the land is considered communal.

Today, expanding access to artificial light sources has made it possible to collect grasshoppers at night with relative ease. Professional collectors can be seen using potent artificial light sources to harvest grasshoppers, although women and children also participate, making use of street illumination (van Huis, 2003b). Some collectors are even charged by electricity companies (US$170 per month) for the provision of constant night-time electricity (Agea *et al.*, 2008). A loss of electricity supply can play havoc on the income earned from harvesting edible grasshoppers (Box 2.13).

BOX 2.13

Power cuts harm Uganda's edible grasshopper business

Power rationing is a common feature in Uganda, with some Kampala households experiencing blackouts lasting more than 48 hours. This has made life difficult for many Ugandan grasshopper catchers and traders.

Julius Kafeero, a grasshopper catcher from the Ugandan capital, Kampala, says electric light is vital for his business. The unreliability of the power supply has forced him and several other collectors to rely on alternative sources of power such as fuel generators.

Despite rising prices, fried grasshoppers remain a delicacy in Uganda. Juliet Nakalyango, a saleswoman at Nakasero market, says customers still buy her grasshoppers, even though prices have doubled. A spoonful of grasshoppers now costs about €0.40 (US$0.50). Last season, the same amount would have bought a whole plastic cupful.

Source: Gitta, 2012.

Commerce

In Uganda, a market study of *Ruspolia nitidula* in Kampala and Makaka districts found that, being a delicacy, 1 kg of grasshoppers fetched prices at local markets that were 40 percent higher than 1 kg of beef (Agea *et al.*, 2008). The study, which interviewed 70 traders and 70 consumers, revealed that retailers bought three-quarters of their supply from wholesalers, and the remainder was derived directly from collectors. The majority of the traders, moreover, indicated that the trade in *R. nitidula* was concentrated along roadsides and/or in service areas along highways. Although men were dominant actors in the trade, women also contributed to collecting. The wholesale price of the grasshoppers was about US$0.56 per kg, while the retail price was about five times higher (US$2.80). On average, traders generated revenues of more than US$200 per season from the sale of *R. nitidula*. One of the issues hindering the sale of *R. nitidula*, however, is that the insect is only seasonally available and shelf life is short.

Other grasshopper species

In Japan, grasshopper harvesting (mainly *Oxya yezoensis*) is connected to the rice harvest. Collection occurs in the morning, when the grasshoppers are wet from the morning dew. The grasshoppers are kept alive for one night after they are collected to allow time for the faeces to be expelled. The next day they are fried or boiled and the legs are removed, as these are not suitable for eating. After being sun-dried, the grasshoppers are cooked in soy sauce and sugar. They are eaten in the autumn as a side dish or snack. Some people store them for up to a year. However, the harvesting and consumption of grasshoppers in Japan has declined in recent years (Nonaka, 2009).

Rice-field grasshoppers are eaten in most Asian countries. In Korea, they were commonly eaten as a side dish, as a lunch-box ingredient and as a snack. The use of rice-field grasshoppers declined during the 1960s and 1970s due to increased insecticide use. In 1981, the rules mandating insecticide use loosened and farmers started using less, which allowed grasshopper populations to increase. The decline in insecticide use and the desire of some Koreans to eat pesticide-free rice led to the development of organic rice farming in Chahwang Myun. This was economically viable because the yields of rice were the same in unsprayed fields as in sprayed fields, and organic rice sold for higher prices. In 1989, the Chahwang Agricultural Cooperative, which functions primarily to buy, mill and sell rice, began to buy dried grasshoppers from farmer-collectors. Three species were present. *Oxya velox* was the most common species (yellow-green and 27–37 mm in length, found in Japan and China and on the Korean Peninsula and the island of Taiwan), comprising 84.5 percent of the total, followed by *Oxya sinuosa* with 14.8 percent and *Acrida lata* with less than 1 percent. In 1991 and 1992, the Chahwang Myun Cooperative continued to buy and sell large numbers of grasshoppers and many people came to buy directly from farmers (Pemberton, 1994).

Forty years ago in Thailand there was an outbreak of the patanga locust (*Patanga succincta*) in maize. Aerial spraying of insecticide did not succeed and a campaign to promote the eating of the patanga locust was initiated between 1978 and 1981. The grasshoppers were deep-fried, used as a cracker ingredient and fermented to make a cooking sauce. Today, the grasshopper (deep-fried) is one of the best-known and most popular edible insects in Thailand, and this species is no longer a major agricultural pest. Some farmers even grow maize crops to feed the insect, rather than harvesting the maize for sale (Hanboonsong, 2010).

The commercialization of grasshoppers is highly dependent on the region. In the Lao People's Democratic Republic, grasshoppers (*Caelifera* spp.) consume weaver ants, the second-best-selling insect on the market (Boulidam, 2010). A number of grasshopper species are collected for family consumption when clearing fields for paddy planting. Cooking is simple: the grasshoppers are salted lightly, boiled in a little water and simmered until dry. Sometimes they are stir-fried, while the bigger ones are deep-fried until crisp, like fried prawns. They can be roasted as well. Usually grasshoppers are served as a single dish and not mixed with vegetables or meat (Chung, 2010).

In Mexico, grasshoppers (*Sphenarium purpurascens*), commonly known as chapulines, are a popular form of street food. Although generally available at informal street stands and small-town restaurants, grasshoppers, among other insects, are now also found on the menus of more expensive restaurants, and dried packaged grasshoppers can be purchased in up-market shops (Ramos Elorduy, 2009).

2.4 IMPORTANT INSECT PRODUCTS

A wealth of bee products – including honey, propolis and beeswax, among others – are well known among the public and are documented extensively by Bradbear (2009). The fact that silk fabric is obtained from silkworms is common knowledge. Yet the general public is less aware of a host of other insect products, many of which are found in most kitchen cabinets, medicines and other household products. Carmine, for example, also called cochineal, is a red dye produced by scale insects, and is typically used to colour food products and as a dye in textiles and pharmaceuticals. Despite its widespread use and approval by the United States Food and Drug Administration, carmine has recently been the subject of controversy, outraging consumers of a popular American coffee chain company because of its use in its beverages (Box 2.14). Silkworm pupae are considered a delicacy in Asia (see section 2.4.2). Lerp (see section 2.4.3) and a host of edible oils derived from pentatomid bugs (see section 2.4.4) are other insect products in common use.

2.4.1 Cochineal

Cochineal (carmine) is a red dye obtained primarily from *Dactylopius coccus* and is used in the food, textile and pharmaceutical industries. The insects live on the cactus *Opuntia ficus-indica*, which is cultivated for its fruits known as prickly pears. The Canary Islands, Chile, Ecuador, Peru and the Plurinational State of Bolivia are the largest

BOX 2.14
Controversial use of cochineal

In early 2012, controversy developed over the Starbucks Coffee Company's strawberry Frappuccino®, after the popular international coffee conglomerate stated that the pink colour of the beverage originated from cochineal extract made from dehydrated and cochineal beetles.

Prior to using cochineal extract, Starbucks had used artificial additives and chose to switch to a more natural means of colouring (Leung, 2012). When this was brought to the attention of a group of vegan[7] consumers in the United States, the story went viral through blogs and web forums and was widely reported in North American media.

In a comment to consumers, the President of Starbucks United States noted that in response to consumer reactions to the use of cochineal extract in some of its products, the company would now use a tomato-based colouring (Burrows, 2012). Within the United States and Canada, the use of the conchineal extract is permitted by their respective food and drug administrations (Health Canada, 2006; USFDA, 2009).

[7] The vegan diet excludes the consumption of animal and animal products, including insects.

producers of cochineal. Between 2000 and 2006, world production increased more than 2.5 times because of increased demand due to growing interest in natural dyes in the food industry (for such products as Campari and Danone strawberry yogurt). In 2006, national production in Peru amounted to 2 300 tonnes (85 percent of global production), with an export value of US$39.6 million. The biggest importers of carmine are Brazil, Denmark, France, Germany and the United States. Other carmine products in Peru include carmine lacquer (US $12.9 million), dried cochineal (US$3.65 million) and carmine acid (US$2.03 million) (Torres, 2008).

In addition to its uses in the food industry, cochineal production has provided a host of social benefits for Peruvians, not least employment. Moreover, production has been hailed for its environmental benefits, as the planting of its host plant, *Opuntia ficus-indica*, protects open spaces from erosion, develops fertile soil for farming, and captures a significant amount of atmospheric carbon.

2.4.2 Silkworm products

Silkworm production is an ancient practice in many parts of Asia, as well as in Europe since its introduction after the crusades. In China, evidence of mulberry production dates back 5 000 years. The famous trade route known as the Silk Road stretched from eastern China to the Mediterranean Sea, making trade in silk, among other products, an international affair as far back as 139 BCE. Silkworm production also has considerable economic relevance, particularly in China and India, where annual production reaches 115 000 and 20 410 tonnes, respectively. More recently, Brazil, Thailand and Uzbekistan have also produced significant quantities.

Aside from mulberry silkworms, significant silk production is obtained from, in decreasing relevance, the Chinese (oak) tussah moth (*Antheraea pernyi*), the camphor silkworm (*Eriogyna pyretorum*), the Thai (or eri) silkworm (*Philosamia* (=*Samia*) *cynthia ricini*) and the Japanese oak silkmoth (*Antheraea yamamai*). Chinese tussah silkworm cocoons production yielded 60 000 tonnes in 2005. Male moths are also used to produce health food products and health wines. In addition, pupae are traditionally eaten and sold in many markets and by vegetable grocers in northeastern China (Zhang, Tang and Chen, 2008). Silkworm pupae are commonly eaten in other Asian countries, including Japan and Thailand and on the Korean Peninsula.

The Thai silkworm is a traditional product that is now distributed on a global scale. The worm is considered a commercially viable product, not only because it produces considerable quantities of silk, but also because its pupae – considered delicacies in China, Japan, Thailand and Viet Nam – are high in protein, making them extremely valid sources of nutrition. About 137 000 households raise silkworms in Thailand, contributing to 80 percent of the country's total silkworm production and to incomes in poor rural households throughout the country. About US$50.8 million was generated from production in 2004 (Sirimungkararat *et al.*, 2010). Pupae of the silkworm are sold processed, packaged and labelled. As such, Thai silkworm pupae can be considered one of the first – if not *the* first – insect product on the global market.

Interest in the possibilities of using silkworms and mulberries for non-textile purposes by making use of silkworm waste was highlighted at the 22nd Conference of the International Sericulture Commission (ISC, 2011). The possibility of silkworm and mulberry production for pharmaceutical and nutritional ends was also explored at the Conference on Sericulture for Multi Products – New Prospects for Development, organized by the Black, Caspian Seas and Central Asia Silk Association (BACSA, 2011). In India, research at the Department of Sericulture at the Tamil Nadul Agricultural University is exploring the possibility of using the high waste streams of the silkworm industry as feed in broiler production (ISC, 2011). In the Republic of Korea, silkworm powder is being produced as a medicine for diabetics because of its blood glucose-lowering effect (Ryu *et al.*, 2012).

2.4.3 Lerp

Lerp is a crystallized sugary secretion produced by the larvae of psyllid insects (belonging to the Hemiptera order) as a protective cover. Psyllids excrete an array of substances because the phloem sap on which they feed is rich in carbohydrates and low in essential nutrients, such as nitrogen; thus, they must suck large amounts of phloem sap to obtain sufficient nutrients and the remainder is excreted as honeydew. The cones of psyllids consist of the insect itself, the secretion and the five exoskeletons the insect sheds when it moults. The conical structures adhere firmly to leaves. Normally the entire insect "cone" is eaten.

Several hundred species of lerp-producing psyllids are found on *Eucalyptus* species in Australia (Yen, 2002). There are also lerp-forming species in Africa and Japan (although possibly only one species in each of these locations). Lerps may have evolved to reduce desiccation in arid environments, and they are an important food source for many birds and mammals. In Australia, for example, the bell miner bird (*Manorina melanophrys*) "farms" psyllid nymphs by removing lerps as a food source but leaving the nymphs, which then construct new lerps (Austin *et al.*, 2004).

The word lerp is derived from the Australian Aboriginal word "larp" for the thick encrustations that form on the insects, which are traditionally collected for food

BOX 2.15

Using scale insects to enhance honey production

The scale insect *Marchalina hellenica* has been introduced in some Mediterranean areas, mainly Greece and Turkey, to increase honey production. The insect sucks the sap of pine trees such as *Pinus brutia*, *P. halepensis*, *P. sylvestris*, *P. nigra* and *P. pinea*. The honeydew the insect produces is an important source of food for honeybees, which produce pine honey. Artificial infestation by beekeepers has resulted in a loss of ecological balance between the insects and their natural predators. As a result, surrounding pine trees are stressed and dying (Gounari, 2006).

(Yen, 2005). Manna, a broader term cited in the Bible and Qu'ran as a "gift of God" that "came from heaven", is believed to denote the same substance, which was found on forest floors, trees and shrubs. While manna has also been used to describe sugary exudations from plants and entire organisms like lichen or fungi (Harrison, 1950), it is also used to refer to manna of animal origin produced indirectly on host plants, including excretions from aphids or coccids, which are insects that feed by sucking up plant juices.

One such "manna" product is mopane bread, which is produced by the psyllid *Arytaina mopane*. These insects feed on the phloem sap of the mopane tree (*Colophospermum mopane*), a common tree species in southern Africa (Sekhwela, 1988). The mopane caterpillar (*Imbrasia belina*) can leave large areas of host trees leafless, and as such is a competitor for lerp production (Hrabar *et al.*, 2009). In nature, animal species often share and compete for food resources. The mopane caterpillar shares its primary food source, the mopane tree, with elephants. Elephants often break stems and branches of mopane trees when feeding and also destroy the plant species preferred by mopane moths for oviposition. Not surprisingly, this elephant activity has been documented to have a negative effect on the abundance of mopane caterpillars (Hrabar *et al.*, 2009). This demonstrates the interdependency of a vertebrate species – the elephant – and two invertebrates – the psyllid and the mopane caterpillar.

Mopane bread delivers 250 calories per 100 g. It has a high percentage of monosaccharides and water-insoluble carbohydrates, low protein content and a high concentration of potassium and phosphorous (Ernst and Sekhwela, 1987), making it a valuable source of nutrition. Mopane bread is only available, however, during the dry season, as rain washes the product from the leaves of trees, although it can be sun-dried and stored. Mopane bread is said to make a delicious meal when mixed with milk (Sekhwela, 1988).

Lerp was particularly popular among Australian Aborigines (Bourne, 1953). Affected leaves were collected and soaked to dissolve the sugar, which acted as a dietary supplement. Yen (2002) described how, in the Australian state of Victoria, lerps were either eaten raw or as a mixture with gum from *Acacia* trees. In arid areas, affected *Eucalyptus* branches were collected and put in the sun and the dried lerps were shaped into balls to be eaten at leisure.

2.4.4 Edible oils from the melon bug and the sorghum bug in the Republic of Sudan

The melon bug (*Coridius (=Aspongopus) vidutus*) is widely distributed throughout the Republic of Sudan, mainly in the western areas of Kordofan and Darfur states, where field watermelons are considered one of the most important crops in traditional rainfed agriculture. Small farmers in these states consider watermelons a strategic crop because of their role as a main source of drinking water during summer and the use of their crop residues as fodder for animals. For this reason, the melon bug is still very much considered a pest – in fact, it is considered the primary pest of watermelons because of the damage it inflicts on watermelon crops. Both nymphs and adults of the bug pierce leaves, stems and young fruits and suck their sap, resulting in wilting, fruit drop and ultimately the death of the plant.

Although the melon bug is considered a pest, its culinary uses are appreciated throughout the country. Melon bugs are generally cooked in their last nymph stage, when they are relatively soft. In Namibia, locals collect the adults and use them as a relish or spice (in powder form). In the western Kordofan state of the Republic of Sudan, the oil extracted from the bugs (after soaking in hot water), known locally as um-buga, is an important source of nutrition. It is also used in cooking in remote areas of the former Sudan and is particularly important when food is scarce. Melon bug oil is used in medicine, for example to cure skin lesions (Mariod, Matthäus and Eichner, 2004).

Besides the nutritional benefits of melon bugs – specifically their oil – the insects boast antibacterial properties. Mustafa, Mariod and Matthäus (2008) tested the oil against seven bacterial isolates and found high antibacterial activity. They concluded that the oil could potentially be used as a preservative in meat and meat products to control gram-positive bacteria (most pathogens in humans are gram-positive). Research has also shown that only very slight chemical changes take place in melon bug oil preserved at temperatures below 30 °C for two years. Furthermore, it was demonstrated that the oxidative stability of sunflower kernel oil was improved by blending it with the highly stable edible oils of the melon bug (as well as the sorghum bug) (Mariod *et al.*, 2005).

The sorghum bug (*Agonoscelis pubescens*) is consumed in Sudan and is known locally as dura (the main pest of sorghum) in both rainfed and irrigated areas. The bugs hibernate from September to December, when they can be found on trees in clusters or in crevices between rocks (van Huis, 2003b). People in the Nuba Mountains in Kurdufan can often be found collecting the insects from these crevices. In western parts of the former Sudan, sorghum bug adults are collected and eaten after frying, and in some areas oil is extracted from the bugs and used for cooking and in medicine. In the Botana area of the former Sudan, nomads use the tar obtained from the bugs after they have been heated to treat their camels for dermatological infections (Mariod, Matthäus and Eichner, 2004). The potential use of these bug oils as biodiesel has been explored, opening insect-related research to an entirely new field (Mariod *et al.*, 2006).

3. Culture, religion and the history of entomophagy

> Disgust is one of our most basic emotions – the only one that we have to learn – and nothing triggers it more reliably than the strange food of others. (Herz, 2012)

3.1 WHY ARE INSECTS NOT EATEN IN WESTERN COUNTRIES?

The Fertile Crescent, a region comprising fertile lands in western Asia and the Nile Valley and Nile Delta in northeast Africa is believed to be one of the regions in which agriculture originated. From there, food production (i.e. plant and animal domestication) spread swiftly throughout Europe (Diamond, 2005). The most valuable wild animal species to be domesticated were large terrestrial mammalian herbivores and omnivores. There are 14 such domesticated mammals worldwide, each weighing at least 45 kg. Remarkably, Eurasia boasted 13 of these animals, and the 14th (the llama) was in the Americas. These animals not only yielded considerable amounts of meat (making them the main providers of animal-based foods), but also warmth, milk products, leather, wool, plough traction and means of transport. It is thought that it was because of the utility of these animals that the use of insects – besides honeybees, silkworms and scale insects – failed to gain much traction in the West. Insects simply could not offer the same benefits. In contrast, the Western Shoshoni of the Central Great Basin in the United States probably relied more on small game (e.g. rodents, lizards and insects) because large game was scarce and did not move in herds (Steward, 1938, cited in Dyson-Hudson and Smith, 1978).

Food production in the Fertile Crescent and Europe led to the domestication of an increasingly wide variety of plants and animals. In turn, agriculture witnessed incredible gains in productivity and efficiency. Food could now be stored, food supplies became more stable, and the hunter-gatherer lifestyle eventually took a back seat to sedentary ways of life that were dependent on farming. This pivotal change in lifestyle combined with the uncertain nature of insects as a staple food because of their seasonality, possibly contributed to the loss of interest in insects as food (DeFoliart, 1999). Although there are records of locusts being consumed in the Fertile Crescent (e.g. Israel) (Amar, 2003), they probably were of minor importance due to the unpredictability of outbreaks.

The importance of sedentary agriculture may have also resulted in the perception of insects as a nuisance and threat to food production. In short, undomesticated food sources in general became less important (DeFoliart, 1999). In modern agriculture, agro-ecosystems are greatly simplified: biodiversity is minimal and the potential to harvest from nature is generally low. Urbanization, which is more extensive in Western countries, has left people out of touch with nature, contrary to many tropical settings where people live a more rural life, although this is changing (UN, 2012). Increasing urbanization will change insect consumption in developing regions of the world if supply to cities remains small and unreliable and urban areas westernize. For example, locust consumption in the Fertile Crescent has disappeared in areas characterized by strong westernization (Amar, 2003).

People in most Western countries view entomophagy with feelings of disgust (Rozin and Fallon, 1987). It is safe to say that most are reluctant to even consider eating insects and, moreover, that they perceive the practice to be associated with primitive behaviour (Vane-Wright, 1991; Ramos Elorduy, 1997; Tommaseo Ponzetta and Paoletti, 1997). Disgust forms a basis of moral judgement and plays a major role in people's rejection of food (Fessler and Navarette, 2003), although it is an innate reaction (Rozin and Vollmecke, 1986; Herz, 2012). Feelings of disgust are mostly triggered by questions such as: *What is it?* or *Where has it been?* (Rozin and Vollmecke, 1986). Aside from basic human emotions, the origins of disgust are rooted in culture (i.e. "taste is culture"), which undoubtedly has a major effect on food habits. Culture, under the influence of environment, history, community structure, human endeavour, mobility and politico-economic systems, defines the rules on what is edible and what is not (Mela, 1999). In short, the acceptance or rejection of entomophagy is a question of culture (Mignon, 2002) (Box 3.1).

BOX 3.1
Sky prawns and sea crickets

Native Americans, such as those who lived freely in what today is called the state of Utah, were very accustomed to eating grasshoppers, locusts and crickets. On their first tasting of shrimp, the Goshute Indians are reported to have named the creatures "sea crickets" (Lockwood, 2004).

Recently in Australia, Christopher Carr and Edward Joshua of the New South Wales Department of Primary Industries proposed the renaming of locusts as "sky prawns", a more acceptable description in Western countries, and compiled recipes in a cookbook, *Cooking with Sky-prawns* (BBC, 2004).

3.1.1 Why are insects consumed in the tropics more than in temperate areas of the world?

It is generally assumed that the practice of eating insects takes place exclusively in tropical countries. This is not entirely true, as insects are also consumed in countries partially or fully in temperate zones, such as China (Feng and Chen, 2003), Japan (Mitsuhashi, 2005) and Mexico (Ramos Elorduy, 1997). Even between and within countries in the tropical zone, there can be large differences among ethnic groups on which insects are considered edible (Meyer-Rochow, 2005). Generally, however, insect consumption is commonplace in the tropics, while in temperate zones it is often absent. A number of trends in favour of entomophagy are recognized in the tropics, although some are admittedly hard to support with literature:

- **Insects tend to be larger in the tropics, which facilitates harvesting.** Although a larger insect body size is often observed in the tropics compared with temperate regions, this trend cannot be generalized (Janzen and Schoener, 1968; Gaston and Chown, 1999). Body size is related to insect metabolism, but how different body sizes occur is not completely clear (Gaston and Chown, 1999). However, nearly all exceptionally large insects are tropical species, and this may be due to some extent to the way in which insects breathe. Like humans, insects require oxygen and produce carbon dioxide (CO_2) as a waste product. Instead of lungs, however, insects use a series of tubes called a tracheal system. The gases are mainly exchanged throughout the body by diffusion, which happens faster at higher temperatures, allowing for the production of bigger insects in warmer climates (Kirkpatrick, 1957). As fossil evidence shows, insects had much larger body sizes during the late Palaeozoic period (Shear and Kukalová-Peck, 1990), some as large as 1 metre, because of higher atmospheric temperatures.
- **In the tropics, insects often congregate in significant numbers, so large quantities can be collected during a single harvest.** Locust swarms settle for the night, making harvesting very easy in the evening and early morning. Winged termites, which take their nuptial flights when the first rains fall after the dry season, emerge from

termite mounds in large numbers. Caterpillars in forests congregate *en masse* by nature. Some insects also congregate in temperate regions, such as the Mormon cricket (*Anabrus simplex*) and the oak processionary caterpillar (*Thaumetopoea processionea*). Native Americans likely ate the Mormon cricket (Madsen and Kirkman, 1988); however, the oak processionary caterpillar has hairs that can cause lepidopterism (dermatitis, conjunctivitis and pulmonary affection) (Gottschling and Meyer, 2006) and should not be eaten.

- **A variety of edible insect species can be found year-round in the tropics.** In temperature zones, insects hibernate to survive cold winters. During this period, no active insect species can be found, and their development comes to a standstill.
- **For many insect species in the tropics, harvests are predictable.** This may not be true for locust swarms, for example, but many locals know where and when to collect a wide range of insect species. Such knowledge has disappeared or is disappearing in temperate and westernizing regions.
- **Location.** Palm weevils, for example, are found in palms that have fallen (e.g. often in typhoons in Asia) or that have been felled deliberately to trigger beetles to lay eggs (Choo, Zent and Simpson, 2009). Bamboo caterpillars can be found in stems of bamboo, dung beetles under dung heaps, soldier termites in termite mounds, and so on. Many insects also have preferred plants or tree species.
- **Time of abundance.** This may be seasonal (often depending on rains) or a preferred time of day. For example, grasshoppers are collected early in the morning when it is too cold for them to fly.

3.2 WHY WERE INSECTS NEVER DOMESTICATED FOR FOOD?

Insects are considered delicacies in many parts of the world, particularly in the tropics. For example, the 1992 *Malawi Cookbook* features many insect-based recipes under the heading "traditional delicacies". It states that a number of insect species are highly sought-after, including barbecued palm weevil larvae and roasted termites. Why then have insect species – with the exception of honeybees, cochineal and silkworms – never been domesticated?

The domestication of animals and plants took place thousands of years ago, with different forms springing up independently in different parts of the world at different times. The Fertile Crescent, China, India, Mesoamerica (central and southern Mexico and adjacent areas), the Andes of South America and the eastern United States all boasted food production from a very early time (Diamond, 2005). A notable case is that of Mesoamerica, where the Aztecs managed to develop a complex society with a high population density without large domesticable animals. One of their prime sources of protein is believed to have been insects and insect eggs, the latter semi-cultivated in marshes and ponds in the Mexican Mesa Central (Parsons, 2010) (see Chapter 4).

Activities surrounding the management of a variety of non-domesticated resources can be witnessed today in tropical forests (Perez, 1995). Recently domesticated tropical plant and animal species include macademia nuts (*Macademia integrifolia*), star fruit (*Averrhoa carambola*), paca (*Agouti paca*) and iguanas (*Iguana iguana*) (Vantomme, Gazza and Lescuyer, 2010). Other instances of the semi-cultivation of edible insects exist (Van Itterbeeck and van Huis, 2012), a well-known example being that of the palm weevil, *Rhynchophorus palmarum*, in Latin America (Choo, Zent and Simpson, 2009). Taming and managing species are stepping stones to domestication (Barker, 2009), yet semi-cultivated and other edible insects – with the exception of honeybees, cochineal and silkworms – were never domesticated. While a simple explanation is not possible here, some important factors can be described.

There are 148 species of large terrestrial mammalian herbivores and omnivores weighing at least 45 kg. The fact that a mere 14 of these have been domesticated is due neither to human ignorance nor human incapacity but is a direct result of the intrinsic biological features of the animals. Diamond (2005) identified six characteristics that a species must have to allow for domestication:

- adequate diet (herbivores are easiest and most efficient to keep as a source of food);
- high growth rate (it is cheaper and more worthwhile to invest in fast-growing animals);
- capacity to breed in captivity (some animals simply refuse to do so);
- a domesticable disposition (e.g. the domestication of horses succeeded but the domestication of zebras failed because of their aggressiveness and tendency to bite relentlessly);
- relatively calm behaviour (animals with tendencies to panic create dangerous situations);
- a clear hierarchical social structure (allowing human to assume the role of leader).

As with mammals, not all edible insect species render themselves to domestication. However, because insects are *not* mammals, the above-mentioned characteristics cannot be assumed to be foolproof in assessing the potential domestication of insect species. Gon and Price (1984) compiled a list of favourable characteristics that can be used to select candidates for insect domestication (these are discussed further in Chapter 7).

The historical contexts in which plant and animal domestication have taken place should also be taken into account. The domestication of large animals (and plants) gave Europeans a considerable advantage over other regions, as evidenced by their worldwide conquests (Diamond, 2005). These conquests enabled Europeans to exert a major influence on food production, with habits, knowledge, techniques and organisms exported worldwide. Perhaps the aforementioned negative attitudes to eating insects formed part of this package, as in more recent times (Box 3.2). It is conceivable that with more time and without European colonialization and imports, the semi-cultivation of edible insects (or even domestication) would be more widespread and involve more species.

BOX 3.2

Examples from Mali and the United States[8]

Western cultures have an embarrassing history of physical, emotional and cultural suppression of indigenous peoples. In 25–50 percent of Native American tribes, for example, there existed a long history of insect eating; yet because Western cultures lacked strong cultural experience with the practice and considered it primitive, they discouraged and suppressed it among Native American tribes when these two cultural groups began to interact in the eighteenth and nineteenth centuries. Western cultures inflicted similar damage on other indigenous groups, including many in sub-Saharan Africa, with the goal of modernizing or westernizing them. This cultural suppression was still prevalant at the end of the twentieth century. As a result, entomophagy has almost disappeared from Canada and the United States and is showing signs of abating in West Africa.

Mali. Traditionally, children in Mali hunt and eat grasshoppers as a snack food. In the village of Sanambele they can be seen harvesting the insects in cotton fields. Since 2010, however, cotton has been grown as a cash crop closer to villages to maintain a high cotton yield in the very fields where children harvest grasshoppers. Western advisors advised farmers to use pesticides to bring more economic stability to the area, an attitude based on zero tolerance for insects in any crop. The fact that grasshoppers form part of this agro-ecosystem and are essential for the nutritional health of the children of Sanambele was not considered. Recent data from Sanambele revealed that 23 percent of these children were already at risk of or had protein energy malnutrition (a condition known as kwashiorkor). Grasshoppers, although a seasonal protein source, supply significant protein to bridge the gap. Mothers in Sanambele, concerned about pesticide

Continues

Box 3.2 continued

exposure, now warn their children not to collect and eat grasshoppers. Western attitudes towards entomophagy have thus resulted in practices detrimental to the people and fragile environments of West Africa.

United States. The Ute, closely related to the eastern Shoshone, are a Native American tribe that lived in what is today Utah in the United States, especially surrounding the Great Salt Lake. In the late 1800s, white settlers arrived from the east in covered wagons, bearing much hope but little or no local or traditional knowledge. Their crops failed, due to low rainfall and grasshopper attack, and it became clear that the reserves of stored food would not sustain the families through the harsh winter. The settlers turned to the Ute to give them food. The Ute prepared their traditional high-protein nutritious snack, called prairie cakes, made from service berries, local nuts and other local materials. The white settlers found them tasty and made it through the winter. Their descendents state that when the settlers later discovered that one of the main ingredients in the prairie cakes was an insect (katydid) that was abundant on the shores of the Great Salt Lake, they refused to consume them – evidence of an existing aversion among Western cultures to food insects 150 years ago. The katydid that saved the lives of these Mormon settlers is now called the Mormon cricket.

[8] This box was contributed by Florence Dunkel.

3.3. NEGATIVE ATTITUDES TOWARDS INSECTS

It is safe to say that, by and large, negative perceptions surrounding insects are fully entrenched in Western societies (Kellert, 1993). Insect harvesting has been associated with the hunter-gatherer era and in turn with "primitive" forms of food acquisition. With the advent of agriculture and the rise of sedentary lifestyles, insects have come to be seen as mere pests (Pimentel *et al.*, 1975; Pimentel, 1991). This is in stark contrast to many tropical regions of the world, where insects have decorative purposes, are used for entertainment and in medicine and sorcery, and are present in myth, legend and dance (Meyer-Rochow, 1979; Yen *et al.*, 2013).

In Western societies – where protein is still largely derived from domesticated animals – insects are virtually synonymous with nuisance: mosquitoes and flies invade homes, the former leaving behind unwanted bites; termites destroy wood possessions; and some insects end up in meals (triggering the disgust factor). Certain insects are also transmitters of disease (Kellert, 1993): a mechanical vector like a housefly, for example, can pick up an infectious agent on the outside of its body and transmit it to food prior to consumption. Biological vectors such as mosquitoes, ticks, fleas and lice harbour pathogens in and are often responsible for serious blood-borne diseases such as malaria, viral encephalitis, Chagas disease, Lyme disease and African sleeping sickness. Arthropods such as spiders have been associated with disease and infection, particularly in Europe, since the tenth century (Davey, 1994). Butterflies and ladybugs are among the few insects that do not evoke aversion, avoidance, disgust and disdain (Kellert, 1993; Looy and Wood, 2006). Few people realize that most insects are beneficial and that very few are damaging.

Western attitudes of disgust towards eating insects have arguably also influenced the preference of people in tropical countries. According to Silow (1983): "It is known that some missionaries have condemned winged termite eating as a heathen custom" and for that reason a Christian person told him that "he would never taste such things, valuing them as highly non-Christian". In Malawi, research has shown that people living in urban areas and devout Christians react with disdain to eating insects (Morris, 2004). As a result of these Western influences, particularly in Africa, research on the contribution of edible insects to nutrition and economy, and into insect species' biology and ecology, has been sporadic (Kenis *et al.*, 2006). Yet insect use in diets may persist, though this may sometimes be admitted only reluctantly by consumers (Tommaseo Ponzetta and Paoletti, 1997). According to DeFoliart (1999), "Westerners should become aware of the fact that their bias against insects as food has an adverse impact, resulting in a gradual reduction in the use of insects without replacement of lost nutrition and other benefits".

However, Western attitudes are changing, as noted by some researchers: "Insects have long been a significant dietary factor in the poorer regions of the world, and it is high time that scientists recognize this fact and begin to build on it, rather than discouraging or ignoring the practice" (Ramos Elorduy, 1990).

3.4 HISTORY OF ENTOMOPHAGY

3.4.1 Entomophagy and religion

Food practices are influenced by culture(s), which have been influenced historically by religious beliefs. The practice of eating insects is cited throughout religious literature in the Christian (Box 3.3), Jewish and Islamic faiths. The Bible speaks of locusts as food in the book of Leviticus, most probably in reference to the desert locust, *Schistocerca gregaria*.[9]

> Yet these may ye eat of every flying creeping thing that goeth upon all four, which have legs above their feet, to leap withal upon the earth (Leviticus XI: 21)

> Even these of them ye may eat; the locust after his kind, and the bald locust after his kind, and the beetle after his kind, and the grasshopper after his kind (Leviticus XI: 22)

BOX 3.3

Entomophagy and modern-day Christianity

In 2012, a Danish priest used entomophagy to demonstrate to his assembly the story of John the Baptist. The New Testament explicitly describes Saint John's source of protein:

> And John was clothed with camel's hair, and with a girdle of a skin about his loins; and he did eat locusts and wild honey (Mark I: 6)

This demonstration was not well received by one of the churchgoers, who complained. According to a Danish bishop, however, the priest had committed no sin because he was demonstrating the word of the Bible. The complainant left the church due to the priest's stunt of eating grasshoppers.

Source: Rohde, 2012.

There are several references in Islamic tradition to insect eating – including locusts, bees, ants, lice and termites (El-Mallakh and El-Mallakh, 1994). The large majority of references are to locusts, specifically mentioning permission to consume the creatures:

> It is permissible to eat locusts (Sahih Muslim, 21.4801)

> Locusts are game of the sea; you may eat them (Sunaan ibn Majah, 4.3222)

> Locusts are Allah's troops, you may eat them (Sunaan ibn Majah, 4.3219, 3220)

[9] Biblical citations provided by Jørgen Eilenberg.

Entomophagy is also present in Jewish literature. Amar (2003) suggested that eating certain species of kosher locusts was largely accepted in ancient times. The practice, however, declined among a considerable part of the Jewish diaspora due to a lack of knowledge about the various types of "winged swarming things" mentioned in the Torah. The tradition was only preserved among Jews of Yemen and in parts of northern Africa. Amar (2003) argued that westernization caused Jews who previously ate locusts to reverse their habits.

3.4.2 Entomophagy in ancient times

The history of entomophagy is well documented by Bodenheimer (1951). In the Middle East, as far back as the eighth century BCE, servants were thought to have carried locusts arranged on sticks to royal banquets in the palace of Asurbanipal (Ninivé). The first reference to entomophagy in Europe was in Greece, where eating cicadas was considered a delicacy. Aristotle (384–322 BCE) wrote in his *Historia Animalium*: "The larva of the cicada on attaining full size in the ground becomes a nymph; then it tastes best, before the husk is broken [i.e. before the last moult]". He also mentioned that, of the adults, females taste best after copulation because they are full of eggs.

References to entomophagy continued throughout the region and the centuries (Box 3.4). In the second century BCE, Diodorus of Sicily called people from Ethiopia *Acridophagi*, or "eaters of locusts and grasshoppers" (*Acrididae* family, Orthoptera order). In Ancient Rome, author, natural philosopher and naturalist Pliny the Elder – author of the encyclopedia *Historia Naturalis* – spoke of cossus, a dish highly coveted by Romans. According to Bodenheimer (1951), cossus is the larva of the longhorn beetle *Cerambyx cerdo*, which lives on oak trees.

Literature from ancient China also cites the practice of entomophagy. Li Shizhen's *Compendium of Materia Medica*, one of the largest and most comprehensive books on Chinese medicine during the Ming Dynasty in China (1368–1644), displays an impressive record of all foods, including a large number of insects. The compendium also highlights the medicinal benefits of the insects.

BOX 3.4
Edible insects through the centuries

Nomads of Arabia and of Libya greet the appearance of locust swarms with joy. They boil and eat them, dry others in the sun and pound them into flour for future consumption – Leo Africanus from Morocco in 1550.

German soldiers in Italy repeatedly and with obvious delight eat fried silkworms – Ulysse Aldovandi in his 1602 treatise, *De Animalibus Insectis Libri Septem*.

We could perhaps in time overcome our repugnance at eating insects and accept them as part of our diet, and then realize that there is nothing terrible about them and that they may perhaps even offer us agreeable sensations. We have grown accustomed to eating frogs, snakes, lizards, shellfish, oysters, etc. in the various provinces of France. Perhaps the first urge to eat them was hunger – René Antoine Ferchault de Réaumur in *Mémoires pour server à l'Histoire des Insectes*, 1737.

Locusts are eaten by most Africans, some Asiatics and especially the Arabs. On their market they appear roasted or grilled in great quantities. When salted, they keep for some time in storage. They are used for supplying ships, when they may be served as dessert or with coffee. This food is in no way repugnant to look at or by association. It tastes like prawn, and is perhaps more delicately flavoured, especially the females when filled with eggs – Foucher d'Obsonville in the 1783 *Philosophic essays on the manners of various foreign animals; with observations on the laws and customs of several eastern nations.*

3.4.3 Modern-day entomophagy

The Italian entomologist and naturalist Ulysse Aldovandi, born in 1522, is considered the founder of the modern-day study of insects. Aldovandi's *De Animalibus Insectis Libri Septem*, published in 1602, is rich in references and concepts derived from his studies as well as original observations. Aldovandi, a specialist in cicadas, suggested that insects were important food items in ancient Far Eastern civilizations, namely China, as far back as several centuries BCE.

Yet it was not until the nineteenth century, when explorers brought back observations from tropical countries, that the Western world grew familiar with the practice of entomophagy. Explorers' accounts of Africa, such as those of David Livingstone and Henry Morton Stanley, which featured stories of insect eating, were instrumental in introducing the practice to the West. In 1857, German explorer Barth Heinrich, for example, wrote in his book *Travels and Discoveries in North and Central Africa* that people who ate insects "enjoy not only the agreeable flavour of the dish, but also take a pleasant revenge on the ravagers of their fields", an interesting take on agricultural pests.

In the United States, swarms of Rocky Mountain locusts (*Melanoplus spretus*) regularly swept across the western half of the country (as far north as Canada) in the nineteenth century, devastating farming communities (Lockwood, 2004). One famed sighting estimated that the locusts spanned 198 000 square miles. This swarm weighed an estimated 27.5 million tonnes and consisted of some 12.5 trillion insects, which according to *The Guinness Book of Records* was the greatest concentration of animals ever recorded.

Leading American entomologist Charles Valentine Riley, appointed in 1868 as the first State Entomologist for the state of Missouri, studied the plague of Rocky Mountain locusts that invaded many western states between 1873 and 1877. He advocated controlling the locusts by simply eating them (Lockwood, 2004):

> Whenever the occasion presented I partook of locusts prepared in different ways, and one day, ate of no other kind of food, and must have consumed, in one form and another, the substance of several thousand half-grown locusts. Commencing the experiments with some misgivings, and fully expecting to have to overcome disagreeable flavor, I was soon most agreeably surprised to find that the insects were quite palatable, in whatever way prepared. The flavor of the raw locust is most strong and disagreeable, but that of the cooked insects is agreeable, and sufficiently mild to be easily neutralized by anything with which they may be mixed, and to admit of easy disguise, according to taste or fancy. But the great point I would make in their favor is that they need no elaborate preparation or seasoning ...

Yet British entomologist V.M. Holt arguably had the most clout in bringing insects to a larger audience through his small booklet published in 1885 titled *Why Not Eat Insects?* The book begged his fellow Englishmen to consider the idea of consuming insects:

> One of the constant questions of the day is, How can the farmer most successfully battle with the insect devourers of his crops? I suggest that these insect devourers should be collected by the poor as food. Why not? (Holt, 1885: 14–15)

Holts' arguments were founded on high moral Victorian values, which included feeding the poor and conserving resources (Friedland, 2007). Holt was greatly puzzled over the lack of acceptance of insects as food, when the composition of other animals that were considered delicacies, like lobster, was nearly the same. However, he did differentiate between insects that he considered unclean and therefore inedible (such as the common fly and the carrion beetle) and clean insects (like cockshafers and grasshoppers). Holt also had an awareness of entomophagy in other cultures:

> If I bring forward examples from ancient times, or from among those nations, in modern times, which are called uncivilized, I foresee that I shall be met with the argument, "Why should we imitate these uncivilized races?" But upon examination it will be found that, though uncivilized, most of these peoples are more particular as to the fitness of their food than we are, and look on us with far greater horror for using, as food, the unclean pig or the raw oyster, than we do upon them for relishing a properly cooked dish of clean-feeding locusts or palm-grubs. (Holt, 1885)

Holding such an opinion in 1885, Holt was clearly ahead of his time and entomophagy was never widely adopted into English food culture.

4. Edible insects as a natural resource

4.1 EDIBLE INSECT ECOLOGY

The edible insect resource is primarily a category of non-wood forest products (NWFPs) collected from natural resources (Boulidam, 2010). Edible insects inhabit a large variety of habitats, such as aquatic ecosystems, forests and agricultural fields. On a smaller scale, edible insects may feed on the foliage of vegetation (e.g. caterpillars) or roots (e.g. witchetty grubs), live on the branches and trunks of trees (e.g. cicadas) or thrive in soils (e.g. dung beetles).

Insect ecology can be defined as the interaction of individual insects and insect communities with the surrounding environment. This involves processes such as nutrient cycling, pollination and migration, as well as population dynamics and climate change. Although more than half of all known living organisms are insects, knowledge of insect ecology is limited. Some species that have long been considered valuable for their products – such as honeybees, silkworms and cochineal insects – are well known, while knowledge of many others remains scarce. This chapter points out the need to study edible insect ecology specifically and shows how this knowledge can be applied.

4.2 Collecting from the wild: potential threats and solutions

4.2.1 Threats

Insects provide essential ecosystem services such as pollination, composting, wildfire protection and pest control (Losey and Vaughan, 2006) (see Chapter 2). Edible insects, such as honeybees, dung beetles and weaver ants, eaten extensively in the tropics, perform many of these ecological services. Until recently, edible insects were a seemingly inexhaustible resource (Schabel, 2006). Yet like most natural resources, some edible insect species are in peril. Ramos Elorduy (2006) identified 14 species of edible insect under threat in Hidalgo state, Mexico, alone, including the red agave worm (*Comadia redtembacheri*) (=*Xyleutes redtembacheri*), which is used in mezcal, the Navajo reservation ant (*Liometopum apiculatum*) and the agave weevil (*Scyphophorus acupunctatus*).

A number of anthropogenic factors impose threats on edible insect populations. Collection itself can result in **direct competition** with other predators, undermining population viability (Choo, 2008). Numerous edible insect species are prey or hosts of other insect species (such as coccinellids and parasitic wasps, respectively) and many other organisms, including birds, spiders, mammals, amphibians, reptiles and fish. The effect of reduced insect populations on their predators is unknown. Many edible insect species are predators themselves or decomposers. A reduction in their numbers may have adverse effects on populations of other insect species and affect ecosystem functions. **Overexploitation** is another serious challenge to both the current and future practice of entomophagy (Morris, 2004; Schabel, 2006), particularly if the number of collected individuals (mature and immature) exceeds regeneration capacity (Cerritos, 2009). In addition, the stability and regeneration of edible insect populations is threatened if collection practices become less selective (Latham, 2003; Illgner and Nel, 2000; Ramos Elorduy, 2006). This happens, for example, when mature insects are collected before their first mating or before they lay eggs (Cerritos, 2009). Moreover, many areas are "open-access", and increased collection efforts could threaten existing populations

(Akpalu, Muchapondwa and Zikhali, 2009) (Box 4.1). Matters are further complicated by the fact that indigenous knowledge – which often includes the sustainable use of edible insects and their habitat – is gradually dissipating (Kenis *et al.*, 2006), and inexperienced collectors sometimes resort to **unsustainable collection methods** (Ramos Elorduy, 2006; Choo, 2008).

BOX 4.1

Lao People's Democratic Republic

In the village of Dong Makkhai in the Lao People's Democratic Republic, 21 species of edible insect are collected and sold at the Sahakone Dan Xang fresh food market. On average, 23 percent of the combined household income of the village is derived from the production and sale of edible insects. Most favoured by consumers are ant "eggs" (larvae and pupae of *Oecophylla smaragdina*), grasshoppers (various species), crickets (*Tarbinskiellus portentosus*, *Teleogryllus mitratus* and *Acheta domesticus*), wasps (*Vespa* spp.), cicadas (*Orientopsaltria* spp.) and honeybees (*Apis* spp.). Today, collectors claim they need more time to find similar amounts of edible insects compared with ten years ago, most likely as a result of the increased number of collectors.

Source: Boulidam, 2010.

Finally, like many other natural resources, **damage to habitat**, such as deforestation, forest degradation and pollution (e.g. through insecticides), has placed further stress on edible insect populations (Morris, 2004; Ramos Elorduy, 2006; Schabel, 2006). Host trees are often cut down to increase and facilitate the collection of insects, such as in the case of the edible caterpillar that feeds on the leaves of the sapele tree, with obvious consequences for future harvests (Vantomme, Göhler and N'Deckere-Ziangba, 2004). Often, habitat damage results from other agricultural activities, such as logging and grazing (FAO, 2004). Effects on insects' habitat invariably influence their abundance and distribution (FAO, 2011c). For this reason, the ways in which **climate change** is likely to affect tropical edible insect populations is still relatively unknown. Increasing temperatures could cause certain populations to increase, although periods of extreme heat or drought could also lead to declines (Toms and Thagwana, 2005). The distribution of species may also be affected.

BOX 4.2

Wild harvesting in Asia and the Pacific: past, present and future

In the past, most wild-harvested insects in Asia and the Pacific were consumed exclusively at the village level, and the quantities collected were determined by personal consumption requirements. Today, wild-harvested insects have become an additional source of income. Where possible, more insects are collected so that a portion of the harvest can be sold at markets and the rest is reserved for personal consumption. With fewer insects available for harvest, fewer are consumed at home, and cash incomes are often used to acquire less-healthy foods. Improved access to insect resources (via new roads and modern transport methods) has resulted in more collectors, often coming from afar. Conversely, this has also enabled local villagers to transport their catches to more distant and larger markets.

Continues

Box 4.2 continued

Increased demand has also caused stress on insect populations and their environment. At present, however, there is a dearth of information on the sustainability of the wild-harvesting of edible insects and its ecological implications. In addition to the adoption of sustainable harvesting methods, other forms of producing edible insects have shown potential for reducing pressure on wild edible insect populations, including habitat management, small-scale controlled rearing in confined conditions (e.g. cages and ponds) and industrial production systems (factories). Species suitable for rearing need to be identified. Making use of the high diversity of invertebrates may help to reduce vulnerability to unforeseen shocks such as disease outbreaks and climatic variability.

Source: Yen, 2012.

The problems faced by edible insect populations relate directly to their collection and are deeply rooted in humankind's unsustainable use of the natural environment. It is possible that collecting edible insects will prove to be a threat to the provision of some essential ecosystem services if the practice is commercialized without attention to their sustainable management (boxes 4.2 and 4.3).

BOX 4.3

Mopane and other African caterpillars

Populations of the mopane caterpillar have dwindled since commercialization got into full swing in the 1990s. Mopane populations have encountered the same problem faced by many NWFPs: once a significant market is found, the pressure to overharvest becomes intense, which generally leads to unsustainable use (Sunderland, Ndoye and Harrison-Sanchez, 2011). Poverty, food insecurity and environmental calamity compound the problem.

In the past, restraint was often used in collecting caterpillars (e.g. the first generation of *Cirina forda* was traditionally left untouched "for the birds" and only the second generation was collected; Latham, 2003). Widespread poverty in rural areas coupled with increasing poverty in urban centres, however, has prompted overharvesting. This has turned the promise of a new source of income and cheaper protein into a conservation dilemma. The overharvesting of mopane populations in Zimbabwe, among other countries, has compromised larval production for years (Roberts, 1998; Illgner and Nel, 2000), and even if environmental conditions were to prove optimal, populations are unlikely to recover. Environmental calamities, such as poor yields brought on by drought, will likely trigger a further increase in harvesting this cheap and generally open-access resource. This has already occurred in many parts of the region (Toms and Thagwana, 2005).

Moreover, when caterpillars cannot be reached, desperate livelihood-seekers will often fell entire trees, a practice traditionally frowned on, as the loss of host trees is detrimental to the survival of future populations (Latham, 2003; Morris, 2004; Toms and Thagwana, 2005). Determining sustainable collection levels remains a difficult task, however, and is a key issue for the future development of the sector, largely because several variables are involved in determining population levels. The seasonality of the resource and unpredictability of outbreaks, coupled with a wide range of complex biotic and environmental factors, among others, make this a contentious issue (Stack *et al.*, 2003; Ghazoul, 2006). These issues deserve research and must be addressed if the sector is to develop sustainably and continue to contribute to lives and livelihoods in southern Africa.

Continues

Box 4.3 continued

Many local communities are well aware of the hazardous practices that can harm the mopane woodlands and are equally conscious of the importance of sound protection measures, including adequate fire management, monitoring the caterpillars and their development, the protection of specific habitats, and adherence to restricted harvesting periods (Holden, 1991; Mbata, Chidumayo and Lwatula, 2002; Toms and Thagwana, 2005). The question is whether these local initiatives are realistic in their socio-economic context. The poverty–environment nexus is not new; economic and nutritional incentives commonly drive local communities to overharvest to meet immediate livelihood needs. Conservation policies need to take this into account. For example, harvesting at the beginning of a season would cause the overexploitation of small larvae, which is a wasteful practice. However, efforts to restrict mopane harvesting periods must provide local people with other options for their diets and livelihoods. Moreover, in some areas, local beliefs do not recognize the lifecycle of the mopane, so habitat management (i.e. restricted harvest) is not understood as a necessity (Toms and Thagwana, 2005). Management measures need to balance ecological as well as social, cultural and economic objectives if they are to have any chance at success.

4.3 CONSERVATION AND MANAGEMENT OF EDIBLE INSECT RESOURCES

Among forest managers, there is little knowledge or appreciation of the potential for managing and harvesting insects sustainably. There is also almost no knowledge or experience in manipulating forest vegetation or harvesting practices to increase, maximize or sustain insect populations. Indeed, because many insects cause massive damage and mortality to valuable commercial trees, many forest managers consider virtually all insects as potential destructive pests. What knowledge does exist with respect to managing insects is often held by traditional forest dwellers and forest-dependent people (Durst and Shono, 2010).

Scientists generally speak about biodiversity on three levels: ecosystem, species and genetic. On all three levels it is believed that biodiversity can make significant contributions to food security and improved nutrition (Toledo and Burlingame, 2006). In view of the ecological services insects provide, deemed vital to human life, the conservation of insects and the habitats they occupy has recently received more attention (DeFoliart, 2005; Samways, 2007). The promotion of "flagship species" is used to stimulate public interest in conservation efforts (Simberloff, 1998). In much the same way, conservation biologists identify "umbrella species" as representative species whose protection is believed to indirectly benefit a large number of naturally co-occurring species and their habitats (Roberge and Angelstam, 2004). While these species tend to be large, emblematic mammals, such as giant pandas and tigers, the possibility of edible insect species as flagship species and/or umbrella species protecting other natural resources deserves attention, not least because of the valuable role they have in the provision of essential ecosystem services (Yen, 2009; DeFoliart, 2005). Yet for this to occur, taxonomic knowledge of insects – which still lags far behind that of vertebrates and plants – needs to improve (Winfree, 2010). There is also relatively little documented knowledge of insect species (compared with vertebrate animal and plant species) about contemporary threats and conservation and management requirements (Yen, 2012). Although insects account for the largest proportion of biodiversity in all forest ecosystems, they continue to be the least studied of forest organisms (Johnson, 2010).

Samways (2007) offered a rare yet promising contribution to insect conservation efforts by describing in detail how insect populations and habitats can be managed and monitored effectively and by identifying the following six principles for maintaining adequate insect population levels: maintain reserves; maintain as much quality landscape heterogeneity as possible; reduce contrast between remnant patches and neighbouring disturbed patches; set aside land for insects outside reserves; simulate natural conditions and disturbance; and connect similar patches of quality habitat. Boulidam (2010) added that efforts in edible insect management ought to focus on edible insect species with the greatest potential and value. These principles are particularly important for forestry, ecology and entomology experts. However, insect conservation efforts will remain futile without adequate support from national and international research and development organizations and local communities (Schabel, 2006; Cerritos, 2009; Boulidam, 2010).

Current gaps in insect ecology are a major impediment to the development and sustainability of entomophagy. Issues that require urgent research include identifying edible insect species, estimating populations, and understanding the ecology and biology of species and their habitats and the factors that determine their abundance. Increased knowledge on factors such as peak abundance, population dynamics and life cycles is essential to counter the depletion of edible insect resources (Ghazoul, 2006; Cerritos, 2009). Tapping into indigenous knowledge is likely to prove particularly useful. In light of the above, obvious next steps in the field of research on edible insects are to sustainably increase the production of wild and reared edible insects, through either expansion or intensification, and to implement ecologically sound forest management practices to this end (Johnson, 2010).

Foresters and forest industries have long considered caterpillars as pests because they feed on fresh leaves (tree foliage) and are therefore perceived to be harmful to tree populations. In reality, however, trees respond to such browsing by producing more foliage. N'Gasse *et al.* (2004) observed that leaf consumption by caterpillars had only a limited impact on trees. In fact, the collection of caterpillars in the forest could be considered a method of biocontrol, so long as trees are not cut during the caterpillar harvest (Vantomme, Göhler and N'Deckere-Ziangba, 2004) (see section 4.3 for more on pest management). In exchange, caterpillar protection could greatly benefit from host tree conservation and management (Holden, 1991; Munthali and Mughogho, 1992; Chidumayo and Mbata, 2002; Toms and Thagwana, 2005).

A red list of endangered insects in East Africa. The International Institute for Tropical Agriculture has established a red list of 34 endangered insect species in Benin. The principal threats to insects involve the deterioration or, in some areas, the disappearance of habitats due to pollution, the overextension of agriculture, poor agricultural practices, uncontrolled burning, the uncontrolled cutting of timber, the disrespect of protected areas and, in the long term, climate change and the disappearance of pollinators. Given that most threatened insects live in forest ecosystems, deforestation is one of the primary concerns (Neuenschwander, Sinsin and Goergen, 2011). The list includes one species that is certainly eaten in Central Africa: the African Goliath beetle (*Goliathus goliathus*) (Bergier, 1941). In Benin, this species is threatened by a reduction in the availability of its preferred host tree, the rare *Holoptelea grandis* (Ulmaceae). Because the Goliath beetle can now be easily reared, hunting for insects has diminished and is no longer considered a threat to these trees (Neuenschwander, Sinsin and Goergen, 2011).

Biodiversity of micro-organisms and invertebrates for food and agriculture. FAO's Commission on Genetic Resources for Food and Agriculture (CGRFA) and the Treaty on Plant Genetic Resources for Food and Agriculture recognize the fundamental role that micro-organisms and invertebrates play in food security and sustainable agriculture through the services they provide in food production ecosystems and the natural environment. Among the functional groups distinguished by the CGRFA are: pollinators; biological control agents (biocontrol); soil ecosystem engineers and regulators; food providers and providers of non-timber forest goods (e.g. silk, honey, edible insects); and aquatic invertebrates and their contribution to fisheries and aquaculture (which could be extended to include the use of invertebrates as feed for conventional livestock) (FAO, 2009a). FAO has a long tradition of technical work on the importance

BOX 4.4
Insects and biodiversity in Brazil

Brazil is recognized globally for its status as a biodiversity hotspot (Myers *et al.*, 2000). The country also hosts a rich cultural diversity, with a remarkable 222 indigenous ethnic groups as well as several other groups, including artisanal fishermen, Amazon caboclos (or river-dwellers) and Afro-Brazilians, also known as quilombolas. This combined diversity is known as biosociodiversity (Costa Neto, 2012).

A total of 135 edible insect species belonging to nine orders (Figure 4.1) and 23 families in 14 of Brazil's 26 states have been documented in literature. Of these, 95 have been identified to the species level and 18 to the genus level, while others are known only by their native names. The most consumed species are in the Hymenoptera (63 percent), Coleoptera (16 percent) and Orthoptera (7 percent) orders (Costa Neto, 2012). Given Brazil's vast biosociodiversity, "it could be stated that anthropoentomophagy is underestimated, since nutritious edible insects are abundantly available" (Costa Neto, 2012).

FIGURE 4.1
Distribution of insects, by order, Brazil

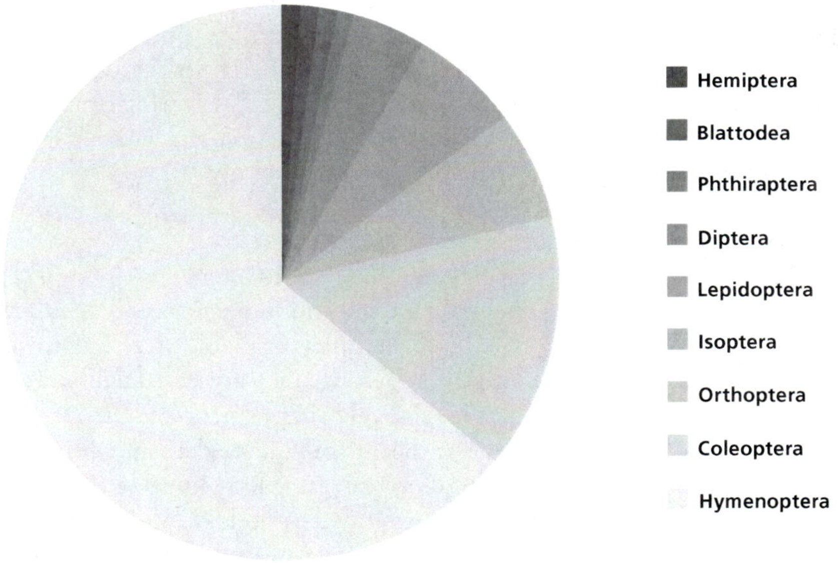

Source: Costa Neto, 2012.

of micro-organisms and invertebrates for food and agriculture. Among these are the Organization's programmes and strategies on integrated pest management (IPM). Through the CGRFA, focus on this "hidden biodiversity" has increased.

FAO also coordinates two global initiatives of the Convention on Biological Diversity, which have been established in recognition of the essential services provided by micro-organisms and invertebrates across all production systems: the International Initiative for the Conservation and Sustainable Use of Pollinators, and the International Initiative for the Conservation and Sustainable Use of Soil Biodiversity. For an example of national insect biodiversity, see Box 4.4. Many partner organizations collaborate with FAO on these important initiatives.

4.4 SEMI-CULTIVATION OF EDIBLE INSECTS

Better knowledge of a particular insect species' biology and ecology can lead to more than an understanding of its seasonality, for example, or the development of efficient tools to collect them. It may also enable the manipulation of an edible insect's habitat at a small or large scale, such as the insects' behaviour and availability throughout the year (Van Itterbeeck and van Huis, 2012). This is termed **semi-cultivation**, as it resembles cultivating – a process that promotes the growth (or quality) of an organism through the use of labour and skill. Semi-cultivation rarely involves the tending of insects (as opposed to actual cultivation, or farming, which is the topic of Chapter 7). Semi-cultivated insects are available in the wild and are generally not grown in captivity (although some may be captive during part of their development, such as the palm weevil larvae, which is farmed as larvae in plastic containers in Venezuela; Cerda *et al.*, 2001). Semi-cultivated insects are thus not isolated from their wild populations. Palm weevil larvae in the Amazon Basin, Indonesia, Malaysia, Papua New Guinea, Thailand and tropical Africa, caterpillars in sub-Saharan Africa, bamboo caterpillars in Thailand, eggs of aquatic Hemiptera in Mexico and termites in sub-Saharan Africa are all examples of semi-cultivated insects, the habitats of which are manipulated to a greater or lesser degree. Evidence of landscape manipulations for food can be found in archaeological records, suggesting that they were key stepping stones in the development of sedentary food production and domestication (Barker, 2009). The manipulations intended for producing edible insects are therefore useful as first steps towards a more controlled production. Semi-cultivation has many benefits, not least of which is ensuring the availability and predictability of edible insects. The activities surrounding semi-cultivation have the potential to contribute to both edible insect habitat conservation and food security. In the tropics, emphasis should be placed on maximizing the productivity of semi-cultivation practices already in place. Such activities could be developed for other edible insect species, provided there is sufficient understanding of their biology and ecology.

4.4.1 Palm weevil larvae

Classic examples of semi-cultivation involve the larvae (grubs) of *Rhynchophorus palmarum* (Central and South America), *R. phoenicis* (Africa) and *R. ferrugineus* (Southeast Asia) (see also section 2.3.2). The most detailed information is available on *R. palmarum*. Palm trees can be considered as *controlled* variables because they are felled deliberately and at a chosen location and time. The process is relatively simple: people return to the trunks one to three months after felling to harvest the larvae (Choo, Zent and Simpson, 2009). Some Amerindians employ this technique as a strategy in long hunting and fishing trips that keep them away from their villages for extended periods (Dufour, 1987).

Amerindians in the Bolivarian Republic of Venezuela are known to use their indigenous knowledge of palm weevils to control the relative quantity of the larvae of two palm weevil species, the South American palm weevil *R. palmarum* and the Bearded weevil *Rhinostomus barbirostris*, the females of which oviposit (lay eggs) in the same trunks. These two weevil species differ in their ovipositing biology: *Rhynchophorus palmarum* adults feed, mate and oviposit on exposed inner palm tissue of felled or naturally fallen palms, whereas *R. barbirostris* oviposit on the intact surface of the trunk and can thus use the entire trunk length to lay eggs. *R. palmarum* adults arrive at the trunks earlier than *R. barbirostris*. Thus, when the former are preferred, softer inner tissue is made available by producing deep cuts in the trunk, making larvae of *R. palmarum* more abundant than those of *R. barbirostris*.

A similar technique is employed with the sago palm (*Metroxylon sagu*) in Papua New Guinea (Mercer, 1994). The mating behaviour of the palm weevil (*R. ferrugineus papuanus*) is gregarious, which easily results in 100 grubs or more in a single trunk. As adult weevils only oviposit in unworked portions of trees that are cut to harvest starch, deliberately felling trees for harvesting larvae increases their numbers. In Papua New Guinea, palm trees that render small quantities of starch are often reserved for the semi-cultivation of palm larvae (Townsend, 1973).

Cerda *et al.* (2001) documented simple cottage methods used by Amerindians in Venezuela to rear the South American palm weevil *R. palmarum* on other crops. The weevil is encouraged to grow on the moriche palm (*Mauritia flexuosa*) because weevils grown on this palm have a protein content of 40 percent on a dry weight basis, which is much higher than those reared on other palm species. Amerindian tribes collect the larvae from the trees after four weeks and continue to rear them for several weeks in plastic containers near home. They feed the larvae with banana pseudostem, vegetable refuse and fruits (Cerda *et al.*, 2001). A pineapple–sugarcane diet has also been explored (Giblin-Davis et al., 1989). Overexploitation was considered to be a potential problem because palm trees are cut in order to trigger egg-laying by the adult beetle and subsequent larval growth. Although Cerda *et al.* (2001) reported that 31 plant species belonging to 12 families can host palm weevils for feeding, careful management of palm trees is still vital to ensure the sustainability of the practice.

Traditional peoples manage and protect some plant species, including palm trees (Politis, 1996), and intervene in the landscape through, for example, forest clearance, planting and plant tending (Barker, 2009). In many parts of the world, trees are felled deliberately to stimulate the production of palm weevil larvae, suggesting that these larvae are a prime insect food source. When trees have been felled to harvest starch, collect fruit, and/or tap the tree for its sap to make wine, the larvae grow in unworked portions of the trunk and have been referred to as byproducts (Dufour, 1987) and second crops (Bodenheimer, 1951). Although dual production may very well be intended, great care should be taken when referring to palm weevil larvae as true byproducts (Van Itterbeeck and van Huis, 2012).

4.4.2 Caterpillars

Box 4.5 presents two examples of habitat management that can increase caterpillar abundance: chitemene shifting cultivation and fire management. As such, these activities can be considered semi-cultivation. Similarly, host tree planting and refraining from cutting down host trees increases egg-laying sites for caterpillars (Takeda, 1990; Latham, 1999). Latham (2003) provided an extensive list of edible caterpillar species and their host plants, and also showed how traditional regulations concerning caterpillar collecting were being maintained or re-instated. Mbata and Chidumayo (2003) described in detail systems in the Kopa area in Zambia in which all levels of the local society are involved: high-density moth egg sites are detected, and the appearance of the first instar and the final instar (the only instar allowed to be collected among the locals) are monitored. Based on these observations, the timing of collection is regulated, and, when necessary, temporal restrictions are imposed on collecting caterpillars. Rituals are also performed and ceremonies held, thus strongly fixing the caterpillars in local culture. In Bas-Congo in the Democratic Republic of the Congo, locals reintroduce caterpillars such as *Cirina forda* on Acacia trees near their houses and allow the caterpillars to grow until they are ready to eat. Some of the caterpillars can be left to pupate. These will develop into adult butterflies, which will lay their eggs in the same area. In this way a caterpillar supply is ensured for the following season (Latham, 2003).

Given the problems facing wild populations – such as irregular and unpredictable outbreaks (Hope *et al.*, 2009) – the domestication of caterpillars and host plants in agroforestry systems is being explored. For example, larvae can be protected from drought, heat and predation – which contribute to mortality in the wild – by using simple techniques such as protective shade cloth sleeves covering branches and shade houses. The long-term viability of such enterprises, however, is unclear. Captive breeding is vulnerable to viral and bacterial diseases and to parasitoids, a problem that also occurs in wild populations of mopane. Recent research suggests that mopane farming should be kept small-scale to reduce the impact of viral diseases. It is better, therefore, for a village to have multiple small-scale farms than one large farm, so that failing farms can easily be restocked with healthy eggs or larvae if information and resources are exchanged readily between farms. Such arrangements cannot be developed overnight, however, as it implies an established degree of trust between farmers. Disease management and control in mopane caterpillar farming deserves serious attention if semi-domestication is to be a serious proposition. The development of captive breeding does not exclude the continued development and success of managing wild populations. On the contrary, further research into reducing mopane mortality from disease would equally benefit wild populations and contribute to the development of a caterpillar-rearing sector (Ghazoul, 2006).

Considering the economic and nutritional importance of mopane caterpillars in Africa, more scientific knowledge relating to wild mopane caterpillar ecology and population biology is necessary. The same is true for a host of other edible caterpillars (Munyuli Bin Mushambanyi, 2000). This is arguably a result of minimal interest in the issues arising from a bias in Western science against insects as a viable source of food and income (Kenis *et al.*, 2006). Filling existing research gaps will provide the knowledge needed for the sustainable management of wild mopane caterpillar populations. Greater accuracy in determining outbreaks, for example, and a better understanding of the affects of diseases and parasitoids on mopane caterpillar populations, would benefit this farming and, in turn, local communities (Ghazoul, 2006).

In Thailand, the Forest Products Research Division of the Royal Forest Department conducted research into the bamboo caterpillar, *Omphisa fuscidentalis*, also known as bamboo worm. The caterpillars reside in bamboo internodes, where they feed on the soft inner tissue. Although the bamboo must be cut and opened to harvest the caterpillars, this does not kill the culm. The bamboo that is inhabited by caterpillars can be identified by the hole made by the young larvae when they enter, as well as by internode size, although amateur gatherers have been known to fell bamboo needlessly in search for caterpillars. Nevertheless, felled bamboo can be used as fuel and garden materials and in the manufacture of handicrafts. The research, including data on the bamboo worm's biology and ecology, was published by the Thai Royal Forest Department in a manual in 2000. Among other things, this manual encourages locals to plant bamboo to compensate for the damage caused by harvesting bamboo caterpillars. Methods are provided for semi-cultivating the bamboo caterpillar: for example, mature larvae can be introduced to bamboo shoots placed in water and covered in a net sleeve at home. The resulting adults then mate, and adult females deposit eggs on the bamboo. Current research is looking into shortening the diapause to allow for year-round production (Singtripop, Wanichacheewa and Sakurai, 2000).

BOX 4.5

Effect of fire management and shifting cultivation on caterpillar populations

Fire management

Fire is often used as a management tool in forestry. In Kasungu National Park, Malawi, where the practice is commonplace, fire policies have reportedly been affecting caterpillar yields (Munthali and Mughogho, 1992). Late burning (from September to October) is particularly hazardous for caterpillar populations as it coincides with the egg-laying period of adult moths. In contrast, early burning (June to July) increases caterpillar yield, probably because it decreases the abundance of predators of moth eggs and larvae and promotes the growth of the young leaves on which caterpillars feed.

It was initially believed that the highest caterpillar yields could be found at tree heights of 1–3 m – coincidentally also within the reach of harvesters. If this is true, a rotation

Continues

Box 4.3 continued

burning scheme promoting growth of stems between 0 m and 4 m might point the way to a sustainable forest management policy in the woodlands, which would benefit lives and livelihoods at the same time (Munthali and Mughogho, 1992). However, a later study suggested that the majority of mopane caterpillars (70 percent) are found above 3 m in host trees (Roberts, 1998). This difference might be explained by differing local conditions in these habitats. Thus, fire management policies should be determined at the local level, and above all they should take into account the important forest resources – such as the mopane caterpillar – that people depend on for food security and livelihoods.

Chitemene shifting cultivation

Chitemene shifting cultivation (i.e. clearing the forest canopy) in fallow lands stimulates the re-growth of caterpillar host trees and thereby increases the potential abundance of caterpillars (Chidumayo and Mbata, 2002). To some extent, selective cutting may favour caterpillars, as it does not appear to negatively affect woodlands. Forest reserves could provide an ideal setting for such forms of habitat management, also providing a forum for educating local communities on the importance of habitat conservation. The role of forest reserves could be reconceptualized to include areas designated for caterpillar production, which would also serve to counteract logging activities on caterpillar host trees (N'Gasse, 2004). However, the long-term effects of such management strategies, as well as the effect on the proportion of tall trees to shorter ones, require further research (Chidumayo and Mbata, 2002).

4.4.3 Eggs of aquatic Hemiptera

The Aztecs considered the eggs of aquatic true bugs as delicacies and referred to them as ahuauhtle; to the Spanish conquistadores they were known as Mexican caviar (although the term is also used for ant larvae) (Bachstez and Aragon, 1945). The ahuauhtle (the adults of which are called axayacatl) measure about 0.5–1 mm. Among the most popular are the eggs of *Corisella*, *Corixa* and *Notonecta* species (Bergier, 1941; Bachstez and Aragon, 1945; Parsons, 2010).

Encouraging the production of ahuauhtle is relatively simply. Female bugs deposit their eggs on aquatic vegetation in lakes (Bergier, 1941; Bachstez and Aragon, 1945; Parsons, 2010). The eggs are semi-cultivated, therefore, by providing egg-laying sites for the female bugs. Bundles of twigs, grasses or reeds (e.g. *Carex* sedges; Guerin-Meneville, 1857) are bound together with a rope and spaced out on the bottom of the lakes (still and shallow water works best), using stones to keep them in place (Guerin-Meneville, 1857; Ramos Elorduy, 1993; Parsons, 2010). Long U-shaped grass/reed bundles, placed at 1 m intervals, have been used more recently. The female Hemiptera lay their eggs on these bundles and can be harvested easily by removing the bundles and shaking them. However, harvests have declined in many of these lakes in recent decades due to pollution (Ramos Elorduy and Pino, 1989).

4.4.4 Termites

Farina, Demey and Hardouin (1991) describe a simple method for semi-cultivating termites involving the replication of interior mound conditions by providing a mix of humidified cellulose (e.g. paper, cardboard and dried plant material) and soil in a cool, dark place. In Togo, local products are used exclusively for the semi-cultivation of termites – an old canari (a plastered water-storage recipient), dry sorghum stems or other cereals, water, a piece of an old jute bag, a stone and some moist soil. A simple wooden construction is made to hold the canari in place over a mound entrance hole; termites can be harvested in 3–4 weeks. Termite mounds that are under construction are ideal, but old mounds also fare well (Farina, Demey and Hardouin, 1991).

4.5. PEST MANAGEMENT

Many edible insects are considered pests and threats to agricultural crops (Box 4.6), and treatment with chemical control methods such as pesticides and insecticides is commonplace in many parts of the world. Manual collection of these pests could not only feed mouths and save crops but also benefit the environment by reducing and mitigating the need for pesticides.

BOX 4.6

The case of the cockchafer bug: from agricultural pest to delicacy to conservation controversy

Though hard to believe, the may bug, and in particular the common European cockchafer bug (*Melolontha melolontha*), has featured as a food item in Europe as far back as the eighteeenth century. Erasmus Darwin (1731–1802), a natural philosopher, physiologist, inventor, poet and grandfather of Charles Darwin, wrote about the may bug as food in *Phytologia* in 1800:

> I have observed the house sparrow destroy the Maychafer, eating out the central part of it, and am told that turkeys and rooks do the same; which I thence conclude might be grateful food, if properly cooked, as the locusts or termites of the East. And probably the large grub, or larva of it, which the rooks pick up in following the plough, is as delicious as the grub called Grugru, and a large caterpillar which feeds on the palm, both of which are roasted and eaten in the West Indies. [The latter probably refers to *Rynchophorus* species, the palm weevil, eaten as a delicacy in virtually all tropical countries.]

Years later (on 13 February 1878) the may bug sparked controversy. In an effort to counter a law calling for the destruction of agricultural pests (in particular the may bug), French senator Tesselin published a recipe for the cockchafer in the *Journal Officiel*:

> Catch the may bugs, pound them and put them through a sieve. For making a thin soup, pour water over them. For making a fat soup, pour bouillon over them. This gives a delightful dish, esteemed by the gourmets.

Remarkably, the may bug became the main ingredient of a traditional soup made in a handful of European countries called cockchafer soup. Until the mid-twentieth century, in fact, this soup – which is comparable with lobster soup – was considered a delicacy in France, Germany and several other European countries. Despite the dish's status as a delicacy, controversy has continued through the years. Today, efforts are being waged to protect the species and its habitat. The case of the cockchafer can be viewed as an encouraging example of the development of edible insects as sources of food because it shows that perceptions can and do change.

Source: Wikipedia, http://de.wikipedia.org/wiki/Maik%C3%A4fersuppe

Cerritos and Cano-Santana (2008) documented the effectiveness of handpicking the popular edible grasshoppers called chapulines (*Sphenarium purpurascens*) from fields of alfalfa to protect the crop and the insect without chemical control (see also section 2.2 on grasshoppers). With comparable (although slightly lower) crop yields, mechanical control has the advantage of considerably lower environmental damage as well as generating an extra source of nutrition and income from the consumption and sale of grasshoppers. Pesticide use is frowned on in many Arab regions; handpicking locusts reduces the impact of the insect on crops and provides an additional source of food. Saeed, Dagga and Sarraf (1993) demonstrated the occurrence of pesticides toxic to humans in locusts that were captured during an outbreak, thereby indicating a health risk to humans consuming these locusts. Cerritos (2009) identified 15 edible insect species considered pests of global or local importance in agro-ecosystems that could be controlled through strategies of alternative management, such as mechanical harvesting, and used widely for human consumption (Table 4.1).

TABLE 4.1
Edible species considered as pests of global or local importance in agro-ecosystems, which could be controlled through strategies of alternative management and used widely for human consumption

Order	Species and common name	Distribution
Orthoptera	*Locusta migratoria*, migratory locust	Intercontinental
	Locustana pardalina, South African migratory locust	Africa
	Schistocerca gregaria, desert locust	Intercontinental
	Zonocerus variegatus, variegated grasshopper	Africa
	Sphenarium purpurascens, chapulines	Mexico
Coleoptera	*Rhynchophorus phoenicis*, African palm weevil	Africa
	Rhynchophorus ferrugineus, Indian red date palm weevil	Asia
	Rhynchophorus palmarum, American palm weevil	America
	Augosoma centaurus, scarab beetle	Africa
	Apriona germari, mulberry longhorn stem beetle	Asia
	Oryctes rhinoceros, coconut rhinoceros beetle	Intercontinental
Lepidoptera	*Agrius convolvuli*, sweet potato hawkmoth	Zimbabwe, South Africa
	Anaphe panda, wild silkmoth	Africa
	Gynanisa maja, emperor moth	Africa

Source: Cerritos, 2009.

A different system of ecologically sound pest management involves weaver ants of the genus *Oecophylla*. This generalist and aggressive predator is a highly efficient biocontrol agent for a variety of commercially important tree species (Peng, Christian and Gibb, 2004). Offenberg and Wiwatwitaya (2009b) showed that there is strong potential to use the weaver ant *Oecophylla smaragdina* both as a food – given the popularity of the brood in countries like Thailand and the Lao People's Democratic Republic (Yhoung-Aree and Viwatpanich, 2005; Sribandit *et al.*, 2008) – and as a biocontrol agent in mango plantations.

4.5.1 Case study: *Oecophylla* spp. weaver ants

The Asian weaver ant, *Oecophylla smaragdina*, is consumed in China, India, Indonesia, the Lao People's Democratic Republic, Myanmar, Papua New Guinea, the Philippines and Thailand (DeFoliart, 2002; Yhoung-Aree and Viwatpanich, 2005; Sribandit *et al.*, 2008). Its African sister species, *O. longinoda*, is consumed in the Democratic Republic of the Congo (DeFoliart, 2002), and in Cameroon a sauce is made based on the worker ants (A. Dejean, personal communication, 2012). The range of the Asian weaver ant, *O. smaragdina*, extends from India to Australia, while the green tree ant, *O. longinoda*, is endemic to tropical Africa. Weaver ants are so-called because they bind ("weave") the leaves of living trees together with silk, secreted by their larvae, to form nests. One ant colony consists of numerous nests, often occupying several trees (Lokkers, 1990).

Typically it is the larvae and pupae ("ant eggs"), particularly the large ones destined to become virgin queens, that are consumed. Adults (workers, virgin queens and males) are less favoured but are used as condiments. Worker ants are added to fish soups in the Lao People's Democratic Republic because of their sour taste, much like lemon is used in many Western countries (J. Van Itterbeeck, personal communication, 2012). Weaver ants also serve as traditional medicine in China and India (Chen and Akre, 1994; Oudhia, 2002) and by Australian Aborigines in northern Australia (Yen, 2005). In Indonesia, the larvae and pupae are used as feed for songbirds and as fishing bait (Césard, 2004a). The Asian weaver ant has been used in mango orchards in Australia's Northern Territory as a biocontrol agent of the mango leafhopper, *Idioscopus nitidulus* (Peng and Christian, 2005) (see Figure 4.2).

FIGURE 4.2
Geographic distribution of *Oecophylla* species

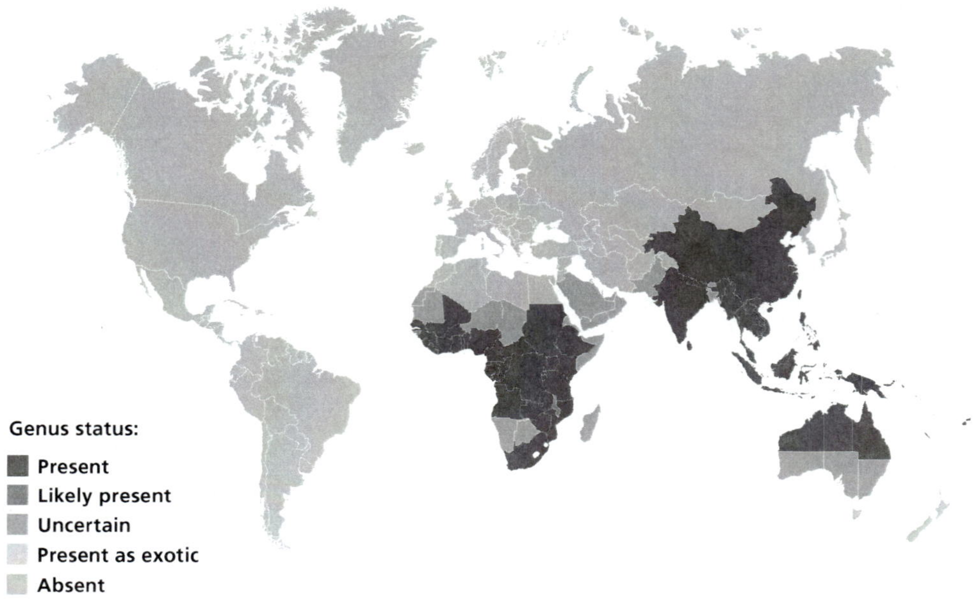

Source: Guénard *et al.*, 2010.

Weaver ants are highly territorial (Hölldobler, 1983) and capture many species of insect that feed on their host trees, which include cashew, cacao, coconut, mango, tea and *Eucalyptus* trees (Peng *et al.*, 2004). As far back as 304 CE they were used in China to protect citrus trees from insect pests. The quality and yield of fruits treated with such biocontrol methods have proved greater than those obtained through conventional insecticide application (Van Mele, 2008). Weaver ants are a perfect example of successful pest management.

Livelihoods

In Thailand and the Lao People's Democratic Republic, the collecting season peaks in February to April and is largely determined by the availability of larvae and pupae destined to become virgin queens (Sribandit *et al.*, 2008; J. Van Itterbeeck, personal communication, 2012). The areas in which ant nests are found are considered open access (Césard, 2004b; J. Offenberg, personal communication, 2010; J. Van Itterbeeck, personal communication, 2012). Larvae and pupae collection, usually carried out by women, provides cash income for many rural people. It is also a valuable source of nutrition: fresh larvae and pupae provide 7 g of protein and 79.2 kilocalories of energy per 100 g (Yhoung-Aree, Puwastien and Attig, 1997). Sribandit *et al.* (2008) estimated that the average household in Thailand consumes 49 kg of larvae and pupae per ant-harvesting season, with the ant harvest constituting some 30 percent of yearly income (and occupying less than 20 percent of labour engagements). This endeavour is therefore an important part of many families' livelihood strategies. Césard (2004b) reported, however, that there are constraints on commercialization due to limited preservation techniques and associated declines in price.

Collection practices
Queen larvae and pupae (ant eggs) are popular foods in Thai urban and rural areas (Yhoung-aree, 2010) and are collected intensively in the latter (Sribandit *et al.*, 2008). Collection is done by penetrating the nest with a long bamboo stick with a basket, bag or net attached to one end to catch the "eggs" inside the nest (Césard, 2004b; Sribandit *et al.*, 2008; J. Van Itterbeeck, personal communication, 2012). "Eggs" do not appear to be overharvested, most probably because population regeneration is quick, for a number of reasons: collectors intentionally refrain from harvesting all larvae and pupae from one nest; the nests of founding queens are usually not collected (e.g. in the Lao People's Democratic Republic, such nests are usually very small and therefore they are ignored because a small yield of ant eggs is anticipated); traditional practice involves rotating between different forest patches over the course of the season; and only a very small fraction of the worker ants are removed (Césard, 2004b; Sribandit *et al.*, 2008; Van Itterbeeck, personal communication, 2012). Nevertheless, care must be taken because, as with most NWFPs, when economic incentives increase (which has already happened in some areas) so too do the number of collectors and, in turn, so does harvesting pressure. Thai collectors have already reported a decline in availability, although it is unsure if this is due to collecting practices, harvesting pressure or forest loss. Moreover, strong competition between collectors risks endangering the continuation of traditional rotation systems and could lead to the adoption of more destructive tools and methods (Sribandit *et al.*, 2008). More positively, it has been observed that when other subsistence activities are present to complement (daily) income, fewer ant eggs tend to be collected (J. Van Itterbeeck, personal communication, 2012).

Weaver ants (*Oecophylla* spp.) can be successful in controlling insect pests, but the aggressive workers are often considered a nuisance, especially when harvesting produce such as fruits. Recent research in Thailand suggests that the use of weaver ants as a biocontrol can be maintained while harvesting the queen larvae and pupae, which are not vital for colony survival. The number of larvae and pupae of the worker caste produced in a colony during the peak ant egg collecting season is believed to be small (Offenberg and Wiwatwitaya, 2009a); these are often not collected in Thailand. Harvested nests even develop higher worker ant densities in time and thus the biocontrol capacity is maintained, if not increased. A new agricultural system could be established in which high fruit yields and quality are achieved and pest insects – that is, weaver ant queen larvae and pupae – are converted into easily manageable and accessible protein foods (Offenberg and Wiwatwitaya, 2009b). In a study in Thailand, Offenberg (2011) identified Asian weaver ants as a very promising insect to farm commercially.

5. Environmental opportunities for insect rearing for food and feed

Feeding a growing world population with more demanding consumers will necessarily require an increase in food production. This will inevitably place heavy pressure on already limited resources such as land, oceans, fertilizers, water and energy. If agricultural production remains in its present form, increases in GHG emissions, as well as deforestation and environmental degradation, are set to continue. These environmental problems, particularly those associated with raising livestock, need urgent attention.

Livestock and fish are important sources of protein in most countries. According to FAO (2006), livestock production accounts for 70 percent of all agricultural land use. With global demand for livestock products expected to more than double between 2000 and 2050 (from 229 million tonnes to 465 million tonnes), meeting this demand will require innovative solutions. Similarly, fish production and consumption has increased dramatically in the last five decades. As a consequence, the aquaculture sector has boomed and now accounts for nearly 50 percent of world fish production. The sustainable growth of the sector will depend largely on the supply of terrestrial and aquatic plant-based proteins for feed. The opportunity for insects to help meet rising demand in meat products and replace fishmeal and fish oil is enormous.

Large-scale livestock and fish production facilities are economically viable because of their high productivity, at least in the short term. However, these facilities incur huge environmental costs (Tilman *et al.*, 2002; Fiala, 2008). Manure, for example, contaminates surface water and groundwater with nutrients, toxins (heavy metals) and pathogens (Tilman *et al.*, 2002; Thorne, 2007). Storing and spreading manure can involve the emission of large quantities of ammonia, which has an acidifying effect on ecosystems. Any increase in animal production will, moreover, require additional feed and cropland and will likely trigger deforestation. The Amazon is a case in point: pasture now accounts for 70 percent of previously forested land, with feed crops covering a large part of the remainder (Steinfeld *et al.*, 2006).

In 2010, Sachs (2010) argued that agriculture was the leading cause of anthropogenic-induced climate change and that the world needed new agricultural technologies and patterns of food consumption based on healthier and more sustainable diets. Feeding future populations will require the development of alternative sources of protein, such as cultured meat, seaweed, beans, fungi and insects.

Consuming insects has a number of advantages:

- They have high feed-conversion efficiency (an animal's capacity to convert feed mass into increased body mass, represented as kg of feed per kg of weight gain).
- They can be reared on organic side streams, reducing environmental contamination, while adding value to waste.
- They emit relatively few GHGs and relatively little ammonia.
- They require significantly less water than cattle rearing.
- They have few animal welfare issues, although the extent to which insects experience pain is largely unknown.
- They pose a low risk of transmitting zoonotic infections.

Despite these benefits, consumer acceptance remains one of the largest barriers to the adoption of insects as viable sources of protein in many Western countries. Nevertheless, history has shown that dietary patterns change quickly, particularly in a globalized world (the rapid acceptance of raw fish in the form of sushi being a good example).

However, replacing a part of conventional meat with edible insects implies an end to unlimited harvesting from nature, as this would place enormous pressure on wild populations. The production of edible insects would need to shift towards rearing either at the cottage-scale level or in large industrial units.

5.1 FEED CONVERSION

As demand for meat rises, so too does the need for grain and protein feeds. This is because far more plant protein is needed for an equivalent amount of animal protein. Pimentel and Pimentel (2003) calculated that for 1 kg of high-quality animal protein, livestock are fed about 6 kg of plant protein. Feed-to-meat conversion rates (how much feed is needed to produce a 1 kg increase in weight) vary widely depending on the class of the animal and the production practices used. Typically, 1 kg of live animal weight in a typical United States production system requires the following amount of feed: 2.5 kg for chicken, 5 kg for pork and 10 kg for beef (Smil, 2002). Insects require far less feed. For example, the production of 1 kg of live animal weight of crickets requires as little as 1.7 kg of feed (Collavo *et al.*, 2005). When these figures are adjusted for edible weight (usually the entire animal cannot be eaten), the advantage of eating insects becomes even greater (van Huis, 2013). Nakagaki and DeFoliart (1991) estimated that up to 80 percent of a cricket is edible and digestible compared with 55 percent for chicken and pigs and 40 percent for cattle. This means that crickets are twice as efficient in converting feed to meat as chicken, at least four times more efficient than pigs, and 12 times more efficient than cattle (see Figure 5.1). This is likely because insects are cold-blooded and do not require feed to maintain body temperature.

FIGURE 5.1
Efficiencies of production of conventional meat and crickets

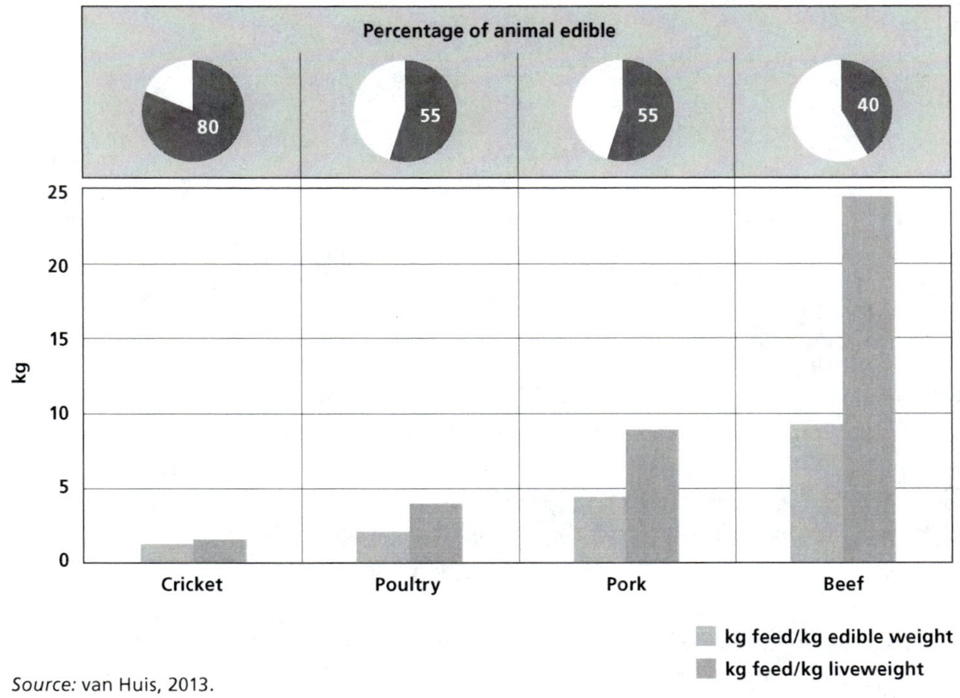

Source: van Huis, 2013.

5.2 ORGANIC SIDE STREAMS

A benefit of insects as an **alternative animal protein source** is that they can be reared sustainably on organic side streams (e.g. manure, pig slurry and compost). The use of organic side streams in insects starts by rearing the insects on biowaste. The insects are processed and fed to a specific animal (Figure 5.2), the meat of which is then sold to the consumer (Veldkamp *et al.*, 2012) (see Chapter 8).

Insect species such as the black soldier fly (*Hermetica illucens*), the common housefly (*Musca domestica*) and the yellow mealworm (*Tenebrio molitor*) are very efficient at bioconverting organic waste. For this reason, these species are receiving increasing attention, as they could collectively convert 1.3 billion tonnes of biowaste per year (Veldkamp *et al.*, 2012). Other insect species, such as crickets, are raised on insect farms and fed with high-quality feed such as chicken feed. The substitution of such feed with organic side streams can help to make insect farming more profitable (Offenberg, 2011). However, at present this is not permitted because of food and feed legislation (see Chapter 14).

FIGURE 5.2
Use of insects in the animal feed chain

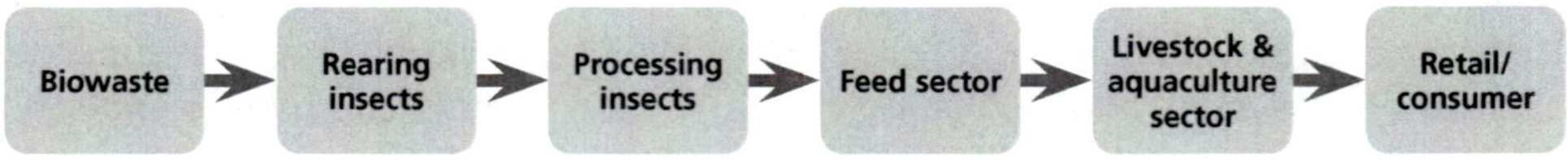

Source: adapted from Veldkamp et al., 2012.

Recycling agricultural and forestry wastes into feed greatly reduces organic pollution. According to DeFoliart (1989), "Practically every substance of organic origin, including cellulose, is fed upon by one or more species of insects, so it is only a matter of time before successful recycling systems will be developed". The possibility of rearing insects on organic waste for human consumption is still being explored, given the unknown risks of pathogens and contaminants (Box 5.1).[10]

BOX 5.1
Ecodiptera project

In 2004, a project co-financed by the European programme LIFE titled Ecodiptera was launched to make better use of the huge volume of pig manure generated across Europe. The overuse of fertilizer is highly linked to a series of environmental problems including nitrification, the excessive enrichment of nutrients in soil and water, and GHGs. Additionally, the use of manure can lead to the spread of pathogens within the environment as well as between humans and animals. In this project, larval flies were used to transform manure into fertilizer and protein. In Slovakia, a pilot plant for the biodegradation of pig slurry was developed with fly larvae by adapting existing technology for chicken manure. Methods suitable for the maintenance of colonies of flies and the identification of optimal conditions were developed. The project found that when flies reach the pupal stage they can be used as protein feed in aquaculture.
The objectives of the project were to:

- demonstrate the technical and economic viability of a new pig manure management method using *Diptera* (maggots);
- obtain a balance between environmental and social concerns in order to increase community acceptance;

Continues

[10] For more information on this topic, refer to the 2012 report, *Insects as a Sustainable Feed Ingredient in Pig and Poultry Diets – A Feasibility Study*, produced by Wageningen University Livestock Research.

Box 5.1 continued

- encourage the progressive phasing out of the current practice of directly using pig manure as an organic fertilizer, which is not recommended due to its high nitrate content;
- prove that the obtained subproducts, such as biodegradable waste remains, pupas and flies, can be included in other processes (e.g. animal feeding and plant pollination) with the goal of obtaining a cycle that produces no waste products;
- introduce a new local juridical model;
- show that fly larvae, previously considered an environmental problem and which occur in pig manure under natural conditions, offer important degradation potential using environmentally friendly management techniques. In this case, the fly problem offers a sustainable solution to the pig manure waste problem.

Source: European Commission, 2008.

5.3.GREENHOUSE GAS AND AMMONIA EMISSIONS

FIGURE 5.3
Relative GHG contributions along the livestock food chain

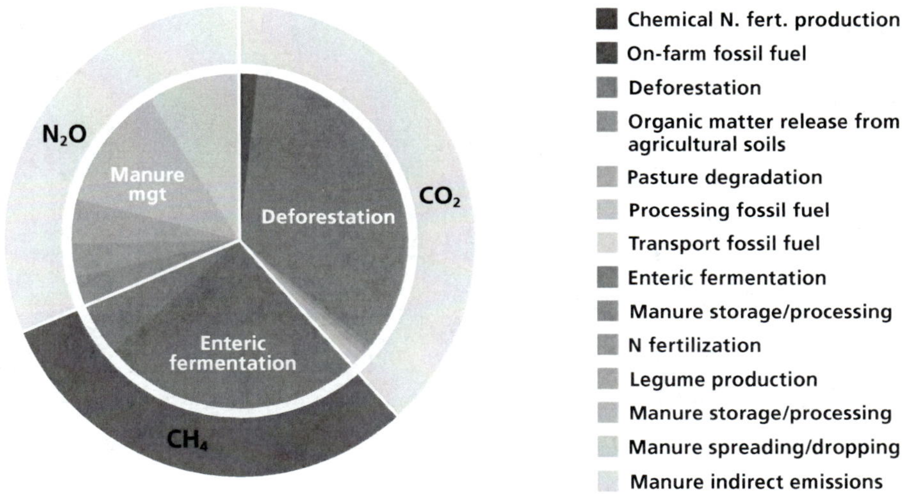

Source: Bonneau, 2008

Livestock rearing is responsible for 18 percent of GHG emissions (CO_2 equivalent), a higher share than the transport sector (Steinfeld *et al.*, 2006). Methane (CH_4) and nitrous oxide (N_2O) have greater global warming potential (GWP) than CO_2: if CO_2 has a value of 1 GWP, CH_4 has a GWP of 23 and N_2O has a GWP of 289 (IPCC, 2007) (Table 5.1).

TABLE 5.1
The animal sector's contribution to GHG emissions

	Carbon dioxide (CO2)	Methane (CH4)	Nitrous oxide (N2O)
Percentage of global emissions	9	35–40	65
Caused by	Fertilizer production for feed crops, on-farm energy expenditures, feed transport, animal product processing, animal transport and land use changes	From enteric fermentation in ruminants and from farm animal manure.	From farm manure and urine

Note: This table shows how much the animal sector contributes to these emissions and why. According to Fiala (2008), 1 kg of beef causes emissions equivalent to 14.8 kg of CO_2, while emissions are lower for pigs and chickens: 3.8 kg and 1.1 kg, respectively.

Source: Steinfeld *et al.*, 2006.

Among insect species, only cockroaches, termites and scarab beetles produce CH_4 (Hackstein and Stumm, 1994), which originates from bacterial fermentation by Methanobacteriaceae in the hindgut (Egert *et al.*, 2003). Yet insects deemed viable for human consumption in the Western world include species such as mealworm larvae, crickets and locusts, which compare favourably with pigs and beef cattle in their GHG emissions (they are lower by a factor of about 100) (Oonincx *et al.*, 2010) (Figure 5.4). Livestock waste (urine and manure) also contributes to environmental pollution (e.g. ammonia) that can lead to nitrification and soil acidification (Aarnink *et al.*, 1995). Mealworm larvae, crickets and locusts also compare favourably to pigs in ammonia emissions, as shown in Figure 5.4 (about a tenfold difference) (Oonincx *et al.*, 2010). These results are taken from small-scale experiments performed in laboratories and caution should be exercised in making comparisons with large-scale pork and beef production.

FIGURE 5.4

Production of GHGs and ammonia per kg of mass gain for three insect species, pigs and beef cattle

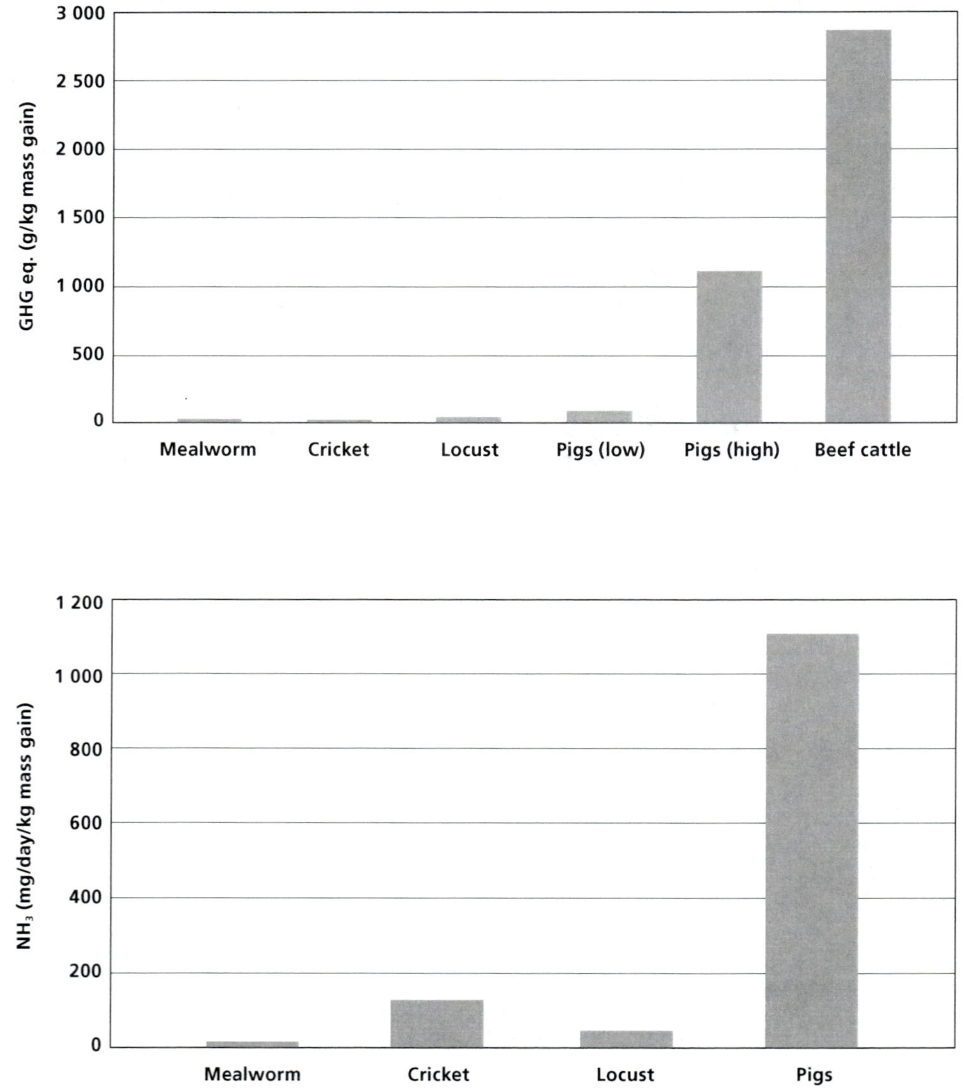

Source: Oonincx *et al.*, 2010.

5.4 WATER USE

Water is a key determinant of land productivity. A growing body of evidence suggests that a lack of water is already constraining agricultural output in many parts of the world. It is estimated that, by 2025, 1.8 billion people will be living in countries or regions with absolute water scarcity, and two-thirds of the world population will likely be under stress (FAO, 2012b). Increasing demands placed on the global water supply threaten biodiversity, food production and other vital human needs. Agriculture consumes about 70 percent of freshwater worldwide (Pimentel *et al.*, 2004). Chapagain and Hoekstra (2003) estimated that producing 1 kg of animal protein requires 5–20 times more water than generating 1 kg of grain protein. This figure approaches 100 times if the water required for forage and grain production is included in the equation (Pimentel and Pimentel, 2003). Chapagain and Hoekstra (2003) described this concept as virtual water. According to the authors, the production of 1 kg of chicken requires 2 300 litres of virtual water, 1 kg of pork requires 3 500 litres and 1 kg of beef requires 22 000 litres, with estimates for the latter reaching as high as 43 000 litres (Pimentel *et al.*, 2004). Estimates of the volume of water required to raise an equivalent weight of edible insects are unavailable but could be considerably lower. Mealworms, for example, are more drought-resistant than cattle (the growing efficiency of mealworms in the presence of sufficient water is described in section 6.1).

5.5. LIFE CYCLE ANALYSIS

Life cycle assessment is a technique to assess the environmental impacts associated with all stages of a product's life, but of the edible insects only mealworms have been assessed in this way. Oonincx and de Boer (2012) quantified GHG production (GWP), energy use and land-use area throughout the mealworm production chain and found that energy usage for the production of 1 kg of mealworm protein was lower than for beef, comparable with pork, and slightly higher than for chicken and milk. GHG emissions due to mealworm production were much lower than for the more common production animals (Figure 5.5). For every 1 ha of land required to produce mealworm protein, 2.5 ha would be required to produce a similar quantity of milk protein, 2–3.5 ha would be required to produce a similar quantity of pork or chicken protein, and 10 ha would be required to produce a similar quantity of beef protein. On the basis of this study, therefore, mealworms are a more environmentally friendly source of animal protein than milk, chicken, pork and beef.

FIGURE 5.5

Greenhouse gas production (global warming potential), energy use and land use due to the production of 1 kg of protein from mealworms, milk, pork, chicken and beef

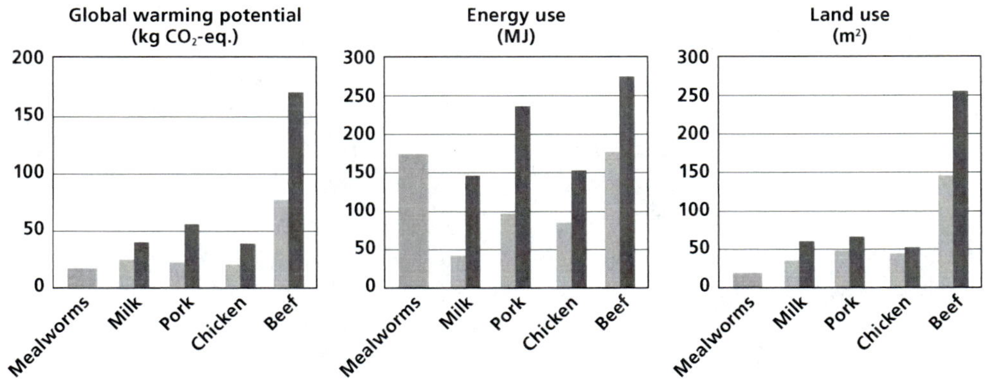

Note: The grey bars are minimal values and the dark green bars are maximum values found in the literature.
Source: Oonincx and de Boer, 2012.

5.6. ANIMAL WELFARE

With respect to intensively farmed animals, Brambell (1965) described the standards that the animal production industry should aspire to: freedom from hunger, thirst, discomfort, pain, injury, disease, fear and distress, and the expression of normal behaviour. Concerning hunger and thirst, this means providing sufficient food of adequate nutrition to prevent, among others, cannibalism. The criteria of freedom from discomfort and the expression of natural behaviour relate to crowding and the tolerance of certain levels of rearing densities. Like many mammals under intensive cultivation, insects are typically reared in small and confined spaces. To ensure animal welfare, farmed insects should be provided with adequate space, which depends on the level of interaction a species has with conspecifics (other organisms of the same species) under natural conditions. For example, locusts reared in captivity are always gregarious and naturally occur in high densities. Mealworms also have a tendency to cluster. In rearing facilities, optimal conditions are pursued to minimize mortality and increase productivity. Little is known about the extent to which insects experience pain and discomfort (Erens *et al.*, 2012), although some research has been carried out using the fruitfly, *Drosophila melanogaster*, as a model organism. Neely *et al.* (2011) looked at nociception, defined as the "sensory perception of potentially damaging noxious stimuli" and found that the genes for nociception were the same for mammals, suggesting that nociception occurs in at least some insects. However, it is uncertain whether these are reflexes or whether higher neural systems are involved. Although there is a lack of evidence that insects possess cognitive ability to experience suffering, some invertebrates, like Cephalopoda, seem to possess advanced cognitive abilities (Crook and Walters, 2011). Until conclusive proof that insects feel pain has been gathered, Eisemann *et al.* (1984) suggested that insects, as a precaution, should be granted the benefit of the doubt. Insect-killing methods that would reduce suffering include freezing or instantaneous techniques such as shredding.

5.7 RISK OF ZOONOTIC INFECTIONS

Intensive animal production with high densities of animals is a starting place for many significant health issues and has been known to trigger the emergence of antimicrobial resistance. Diseases are the cause of large-scale losses of animals, either through high mortality or because of culling policies. Some of these diseases are zoonotic (e.g. H5N1, the highly pathogenic avian influenza, foot and mouth disease, bovine spongiform encephalopathy and Q-fever).

A **zoonosis** is an infection or infestation shared in nature by humans and wild or domestic animals. Recently there seems to have been a serious increase in zoonotic diseases due to the intensification of animal production and climate change. In recent years, the emergence of severe acute respiratory syndrome coronavirus (known as SARS) and influenza A viruses (H5N1 and H7N7) has caused global concern about the potential for pandemics. Many past zoonoses have remained within confined populations; however, in a globalized world the likelihood of such pandemics is increasing. A number of examples exist from various parts of the world, including cutaneous zoonotic leishmaniasis in Manaus, Brazil; ebola, monkeypox and Rift Valley fever in Africa and the Arabic Peninsula; Crimea Congo haemorrhagic fever in the Middle East; bovine spongiform encephalopathy (BSE) in Europe and elsewhere; West Nile fever in Canada and the United States; and paramyxoviruses in Australasia. These demonstrate that a wide variety of animal species, both domesticated and wild, act as reservoirs for pathogens, which may take the form of viruses, bacteria or parasites (Meslin and Formenty, 2004).

In the livestock sector, pathogens that carry infectious diseases are subjected to pressures resulting from the production, processing and retail environment. Taken together, these alter host contact rates, population size and/or microbial traffic flows in the food chain. Insects for food and feed have not been tested sufficiently to determine the risk that they will transmit diseases to humans. Intensive insect-rearing facilities will also be subject to the same pressures exhibited in animal production and, at present, it is not fully understood whether they could be a source of harmful emerging pathogens. Particular attention should be paid to pathogens that initially have animals as hosts but may shift to humans as their preferred hosts. Some well-known diseases (e.g. HIV-related diseases) have been introduced by animals in this way. Pathogen transmission occurs first by the adaptation of the pathogens to the new host population and second by spreading inside the host population. The adaptation of the pathogen to the new host is dependent on the genetic difference between the two species and the nature of the pathogen itself (Slingenbergh *et al.*, 2004).

Because insects are taxonomically much more distant from humans than conventional livestock, the risk of zoonotic infections is expected to be low. Nevertheless, insects are potential vectors of medically relevant pathogens, including the eggs of gastrointestinal helminths found in human faeces. The risk of zoonotic infections (transmitting diseases from humans to animals and back) could rise with the careless use of waste products, the unhygienic handling of insects, and direct contact between farmed insects and insects outside the farm due to weak biosecurity. More research in this area is needed. Safety issues and the hygienic handling of insects are discussed in Chapter 10.

5.8 THE ONE HEALTH CONCEPT

FAO, the World Health Organization (WHO) and the World Organisation for Animal Health use the following informal definition of One Health: "the collaborative efforts of multiple disciplines working locally, nationally and globally to attain optimal health for people, animals and our environment". One Health is a means of managing the threats posed by the interface between human, animal and environmental health. This view of health acknowledges the strong linkages and interconnectedness between the animal–human–environmental health nexus. The health approach is coupled with the biosecurity approach: "Biosecurity is a strategic and integrated approach to analysing and managing relevant risks to human, animal and plant life and health and associated risks to the environment" (WHO/FAO, 2010). Entomophagy, as well as the use of insects for feed, is an area in which the application of the One Health concept would be appropriate, although it requires more research.

6. Nutritional value of insects for human consumption

6.1 NUTRITIONAL COMPOSITION

The nutritional values of edible insects are highly variable, not least because of the wide variety of species. Even within the same group of edible insect species, values may differ depending on the metamorphic stage of the insect (in particular, for species with a complete metamorphosis – known as holometabolous species – such as ants, bees and beetles), and their habitat and diet. Like most foods, preparation and processing methods (e.g. drying, boiling or frying) applied before consumption will also influence nutritional composition. A few scattered studies analyse the nutritional value of edible insects; however, these data are not always comparable due to the above-mentioned variations between insects and because of the varying methodologies employed to analyse the compounds. Moreover, where commonly consumed, insects comprise only a part of local diets. For example, in certain African communities insects form 5–10 percent of the protein consumed (Ayieko and Oriaro, 2008). Nevertheless, because of their nutritional value they are still a highly significant food source for human populations. Attempts are now being made to compile data on the nutritional value of insects (Box 6.1).

This chapter looks at nutritional aspects of insects for human consumption, while Chapter 8 touches on insects in relation to animal nutrition. The main components of insects are protein, fat and fibre; nutritional values are expressed in this chapter as dietary energy, proteins, fatty acids, fibres, dietary minerals and vitamins.

BOX 6.1

The FAO/INFOODS food composition database for biodiversity

The International Network of Food Data Systems (INFOODS), established in 1984, aims to stimulate and coordinate efforts to improve the quality and worldwide availability of food analysis data and to ensure that all people in different parts of the world can obtain adequate and reliable food composition data. INFOODS and FAO are collecting data on food composition and consumption at many levels (e.g. variety, cultivar and breed), and on wild and underused foods in order to promote biodiversity. The first version of the INFOODS Food Composition Database for Biodiversity, comprising analytical data from published and unpublished literature, was launched on 15 December 2010 and now includes the nutritional values of certain edible insects. To be included, nutritional values must be expressed as a 100 g edible portion on a fresh weight basis (FAO, 2012f).

Rumpold and Schlüter (2013) compiled nutrient compositions for 236 edible insects, as published in the literature (based on dry matter). Although significant variation was found in the data, many edible insects provide satisfactory amounts of energy and protein, meet amino acid requirements for humans, are high in monounsaturated and/or polyunsaturated fatty acids, and are rich in micronutrients such as copper, iron, magnesium, manganese, phosphorous, selenium and zinc, as well as riboflavin, pantothenic acid, biotin and, in some cases, folic acid.

6.1.1 Dietary energy

Ramos Elorduy *et al.* (1997) analysed 78 insect species from Oaxaca state, Mexico, and determined that caloric content was 293–762 kilocalories per 100 g of dry matter. For example, the gross energy (which is normally higher than metabolizable energy) of the migratory locust (*Locusta migratoria*) was in the range 598–816 kJ per 100 g fresh weight (recalculated from dry matter), depending on the insect's diet (Oonincx and van der Poel, 2011). Table 6.1 presents energy values expressed in kilocalories per 100 g fresh weight of selected wild and farmed insects worldwide.

TABLE 6.1
Examples of energy content of differently processed insect species, by region

Location	Common name	Scientific name	Energy content (kcal/100 g fresh weight)
Australia	Australian plague locust, raw	*Chortoicetes terminifera*	499
Australia	Green (weaver) ant, raw	*Oecophylla smaragdina*	1 272
Canada, Quebec	Red-legged grasshopper, whole, raw	*Melanoplus femurrubrum*	160
United States, Illinois	Yellow mealworm, larva, raw	*Tenebrio molitor*	206
United States, Illinois	Yellow mealworm, adult, raw	*Tenebrio molitor*	138
Ivory Coast	Termite, adult, dewinged, dried, flour	*Macrotermes subhyalinus*	535
Mexico, Veracruz State	Leaf-cutter ant, adult, raw	*Atta mexicana*	404
Mexico, Hidalgo State	Honey ant, adult, raw	*Myrmecocystus melliger*	116
Thailand	Field cricket, raw	*Gryllus bimaculatus*	120
Thailand	Giant water bug, raw	*Lethocerus indicus*	165
Thailand	Rice grasshopper, raw	*Oxya japonica*	149
Thailand	Grasshopper, raw	*Cyrtacanthacris tatarica*	89
Thailand	Domesticated silkworm, pupa, raw	*Bombyx mori*	94
The Netherlands	Migratory locust, adult, raw	*Locusta migratoria*	179

Source: FAO, 2012f.

6.1.2 Protein

General information about proteins and various amino acids, as well as protein quality, are provided in Box 6.2.

BOX 6.2
Proteins and amino acids ("food chemistry")

Proteins are organic compounds consisting of amino acids. They are important elements of food nutrition but also contribute to its physical and sensory properties. The nutritive value depends on several factors: **protein content**, which varies widely among all foods; **protein quality,** which depends on the kind of amino acids present (essential or non-essential) and whether the quality complies with human needs; and **protein digestibility,** which refers to the digestibility of the amino acids present in the food.

Amino acids are the building blocks required for the biosynthesis of all proteins through human metabolism to ensure proper growth, development and maintenance.

Essential amino acids are indispensable because the body cannot synthesize them and so must obtain them through food. Eight amino acids are classified as essential: phenylalanine, valine, threonine, tryptophan, isoleucine, methionine, leucine and lysine.

Xiaoming *et al.* (2010) evaluated the protein content of 100 species from a number of insect orders. Table 6.2 shows that protein content was in the range 13–77 percent of dry matter and that there was large variation between and within insect orders.

TABLE 6.2
Crude protein content, by insect order

Insect order	Stage	Range (% protein)
Coleoptera	Adults and larvae	23 – 66
Lepidoptera	Pupae and larvae	14 – 68
Hemiptera	Adults and larvae	42 – 74
Homoptera	Adults, larvae and eggs	45 – 57
Hymenoptera	Adults, pupae, larvae and eggs	13 – 77
Odonata	Adults and naiad	46 – 65
Orthoptera	Adults and nymph	23 – 65

Source: Xiaoming *et al.*, 2010.

Bukkens (1997) showed that the mopane caterpillar had lower protein content when dry-roasted than when dried (48 and 57 percent, respectively). The same was true for termites: protein content was 20 percent in raw termites and 32 percent and 37 percent of fresh weight when fried and smoked, respectively (the difference due to varying water content). Protein content is high in insects and therefore using insects as food can help increase dietary quality when including animal source proteins.

TABLE 6.3
Comparison of average protein content among insects, reptiles, fish and mammals

Animal group	Species and common name	Edible product	Protein content (g/100 g fresh weight)
Insects (raw)	Locusts and grasshoppers: *Locusta migratoria, Acridium melanorhodon, Ruspolia differens*	larva	14–18
	Locusts and grasshoppers: *Locusta migratoria, Acridium melanorhodon, Ruspolia differens*	Adult	13–28
	Sphenarium purpurascens (chapulines – Mexico)	Adult	35–48
	Silkworm (*Bombyx mori*)	Caterpillar	10–17
	Palmworm beetles: *Rhynchophorus palmarum, R. phoenicis, Callipogon barbatus*	Larva	7–36
	Yellow mealworm (*Tenebrio molitor*)	Larva	14–25
	Crickets	Adult	8–25
	Termites	Adult	13–28
Cattle		Beef (raw)	19–26
Reptiles (cooked)	Turtles: *Chelodina rugosa, Chelonia depressa*	Flesh	25–27
		Intestine	18
		Liver	11
		Heart	17–23
		Liver	12-27
Fish (raw)	Finfish	Tilapia	16–19
		Mackerel	16–28
		Catfish	17–28
	Crustaceans	Lobster	17–19
		Prawn (Malaysia)	16–19
		Shrimp	13–27
	Molluscs	Cuttlefish, squid	15–18

Source: FAO, 2012f.

The protein content of insects also varies strongly by species. As shown in Table 6.3, some insects compare favourably with mammals, reptiles and fish.

Protein content also depends on the feed (e.g. vegetables, grains or waste). Grasshoppers in Nigeria that are fed with bran, which contains high levels of essential fatty acids, have almost double the protein content of those fed on maize. The protein content of insects also depends on the metamorphosis stage (Ademolu *et al.*, 2010): adults usually have higher protein content than instars (Table 6.4).

TABLE 6.4
Variation in insect protein along subsequent metamorphosis phases of the variegated grasshopper, *Zonocerus variegatus* (raw), Ogun state, Nigeria

Insect stage	Gram protein/100 g fresh weight
Instar:	
First	18.3
Second	14.4
Third	16.8
Fourth	15.5
Fifth	14.6
Sixth	16.1
Adult	21.4

Source: Ademolu, Idowu and Olatunde, 2010.

In Mexico, the protein content of 78 evaluated species ranged from 15 percent to 81 percent of dry matter and protein digestibility ranged from 76 percent to 98 percent (Ramos Elorduy *et al.*, 1997). Comparable studies have been conducted on single species, such as the mopane caterpillar (Headings and Rahnema, 2002) and the field cricket *Gryllus testaceus* (Wang *et al.*, 2004). Bukkens (2005) analysed the protein content of 17 caterpillar species of the family Saturniidae (of which the mopane caterpillar is a member) and found protein content in the range 52–80 percent of dry matter.

6.1.3 Amino acids

Cereal proteins that are key staples in diets around the world are often low in lysine and, in some cases, lack the amino acids tryptophan (e.g. maize) and threonine. In some insect species, these amino acids are very well represented (Bukkens, 2005). For example, several caterpillars of the Saturniidae family, palm weevil larvae and aquatic insects have amino acid scores for lysine higher than 100 mg amino acid per 100 g crude protein. Yet in order to make recommendations regarding the use of edible insects as food enrichments in diets, it is important to look at traditional diets in their entirety, and in particular at staple foods, and to compare their nutritional quality against that of edible insects locally available in the region. In the Democratic Republic of the Congo, for example, lysine-rich caterpillars complement lysine-poor staple proteins. Likewise, people in Papua New Guinea eat tubers that are poor in lysine and leucine but compensate for this nutritional gap by eating palm weevil larvae. The tubers provide tryptophan and aromatic amino acids, which are limited in palm weevils (Bukkens, 2005). In countries in Africa where maize is a staple food – such as Angola, Kenya, Nigeria and Zimbabwe – there are occasionally widespread tryptophan and lysine deficiencies; supplementing diets with termite species like *Macrotermes bellicosus* (Angola) should be a relatively easy step, as they already form accepted parts of traditional diets. Not all termite species are suitable, however: *Macrotermes subhyalinus*, for example, is not rich in these amino acids (Sogbesan and Ugwumba, 2008).

6.1.4 Fat content

Fat is the most energy-dense macronutrient in food. It consists of triglycerides, which all have a glycerol molecule and three fatty acids in their molecular makeup. Box 6.3 provides information on saturated, unsaturated and essential fatty acids.

BOX 6.3
Fatty acids

Saturated fatty acids. In general, saturated fatty acids have a higher melting point than unsaturated fatty acids and are solid at room temperature. They are often found in animal products and tropical oils (e.g. palm and coconut oil).

Unsaturated fatty acids. These consist of mono-unsaturated fatty acids and polyunsaturated fatty acids and are generally liquid at room temperature. Unsaturated fats consist of at least one double bond, and yield slightly less energy during metabolism. They are mostly present in vegetable oils, nuts and seafood. Unsaturated fatty acids are considered better for human health than saturated fat.

Essential fatty acids. These cannot be synthesized by the human body, which means that they must be obtained from the diet. They include some omega-3 fatty acids (e.g. α-linolenic acid) and some omega-6 fatty acids (e.g. linoleic acid).

An example of an edible insect species with high fat content (38 percent of dry weight) is Australia's witchetty grub (Box 6.4). These are very rich in oleic acid, which is an omega-9 mono-unsaturated fatty acid (Naughton, Odea and Sinclair, 1986).

BOX 6.4
Witchetty grub

Witchetty (also spelt witjuti) grubs refer to the large, white, wood-eating larvae of several moths (Cossidae and Hepialidae) and beetles (Cerambycidae) found in Australia. However, the term applies mostly to the larva of the cossid moth, *Xyleutes* species, which can be found 60 cm below ground feeding on the roots of river red gums (*Eucalyptus camaldulensis*). The grub is the most important insect food of the desert and was a staple in the diets of Aboriginal women and children. Edible either raw or lightly cooked in hot ashes, they are sought by Aborigines as a high-protein, high-fat food. The raw witchetty grub tastes like almonds; when cooked, the skin becomes crisp like roast chicken and the inside becomes light yellow in colour.

Edible insects are a considerable source of fat. Womeni *et al.* (2009) investigated the content and composition of oils extracted from several insects (see Table 6.5). Their oils are rich in polyunsaturated fatty acids and frequently contain the essential linoleic and α-linolenic acids. The nutritional importance of these two essential fatty acids is well recognized, mainly for the healthy development of children and infants (Michaelsen *et al.*, 2009). Greater attention has been paid to the potential deficient intake of these omega-3 and omega-6 fatty acids in recent times, and insects could play an important role, in particular in landlocked developing countries with lower access to fish food sources, by supplying these essential fatty acids to local diets (N. Roos, personal communication, 2012). The fatty acid composition of insects appears to be influenced by the plants on which they feed (Bukkens, 2005). The presence of unsaturated fatty acids will also give rise to rapid oxidation of insect food products during processing, causing them to go rancid quickly.

TABLE 6.5
Fat content and randomly selected fatty acids of several edible insect species consumed in Cameroon

Edible insect species	Fat content (% of dry matter)	Composition of main fatty acids (% of oil content)	SFA, MUFA or PUFA1
African palm weevil (*Rhynchophorus phoenicis*)	54%	Palmitoleic acid (38%)	MUFA
		Linoleic acid (45%)	PUFA
Edible grasshopper (*Ruspolia differens*)	67%	Palmitoleic acid (28%)	MUFA
		Linoleic acid (46%)	PUFA
		α-Linolenic acid (16%)	PUFA
Variegated grasshopper (*Zonocerus variegates*)	9%	Palmitoleic acid (24%)	MUFA
		Oleic acid (11%)	MUFA
		Linoleic acid (21%)	PUFA
		α-Linolenic acid (15%)	PUFA
		γ-Linolenic acid (23%)	PUFA
Termites (*Macrotermes* sp.)	49%	Palmitic acid (30%)	SFA
		Oleic acid (48%)	MUFA
		Stearic acid (9%)	SFA
Saturniid caterpillar (*Imbrasia* sp.)	24%	Palmitic acid (8%)	SFA
		Oleic acid (9%)	MUFA
		Linoleic acid (7%)	PUFA
		α-Linolenic acid (38%)	PUFA

Note: 1SFA – saturated fatty acids; MUFA and PUFA – mono and poly unsaturated fatty acids.
Source: Womeni *et al.*, 2009.

6.1.5 Micronutrients

Micronutrients – including minerals and vitamins – play an important role in the nutritional value of food. Micronutrient deficiencies, which are commonplace in many developing countries, can have major adverse health consequences, contributing to impairments in growth, immune function, mental and physical development and reproductive outcomes that cannot always be reversed by nutrition interventions (FAO, 2011c). In insects, metamorphic stage and diet highly influence nutritional value, making all-encompassing statements on the micronutrient content of insect species of little value. Moreover, the mineral and vitamin contents of edible insects described in the literature are highly variable across species and orders. Consumption of the entire insect body generally elevates nutritional content. A study on small fish, for example, suggested that consuming the whole organism – including all tissues – is a better source of minerals and vitamins than the consumption of fish fillets. In much the same way, consuming the entire insect is expected to provide higher micronutrient content than eating individual insect parts (N. Roos, personal communication, 2012).

6.1.6 Minerals

Minerals play an important part in biological processes. The recommended dietary allowance (RDA) and adequate intake are generally used to quantify suggested daily intake of minerals. Table 6.6 compares the RDA of minerals for a 25-year-old male with those provided by the mopane caterpillar. From the table, it is clear that the mopane caterpillar – like many edible insects – is an excellent source of iron. Most edible insects boast equal or higher iron contents than beef (Bukkens, 2005). Beef has an iron content of 6 mg per 100 g of dry weight, while the iron content of the mopane capterpillar, for example, is 31–77 mg per 100 g. The iron content of locusts (*Locusta migratoria*) varies between 8 and 20 mg per 100 g of dry weight, depending on their diet (Oonincx *et al.*, 2010).

Edible insects are undeniably rich sources of iron and their inclusion in the daily diet could improve iron status and help prevent anaemia in developing countries. WHO has flagged iron deficiency as the world's most common and widespread nutritional disorder. In developing countries, one in two pregnant women and about 40 percent of preschool children are believed to be anaemic. Health consequences include poor pregnancy outcomes, impaired physical and cognitive development, increased risk of morbidity in children and reduced work productivity in adults. Anaemia is a preventable deficiency but contributes to 20 percent of all maternal deaths. Given the high iron content of several insect species, further evaluation of more edible insect species is warranted (FAO/WHO, 2001b).

Zinc deficiency is another core public health problem, especially for child and maternal health. Zinc deficiencies can lead to growth retardation, delayed sexual and bone maturation, skin lesions, diarrhea, alopecia, impaired appetite and increased susceptibility to infections mediated via defects in the immune system (FAO/WHO, 2001b). In general, most insects are believed to be good sources of zinc. Beef averages 12.5 mg per 100 g of dry weight, while the palm weevil larvae (*Rhynchophorus phoenicis*), for example, contains 26.5 mg per 100 g (Bukkens, 2005).

TABLE 6.6
Recommended intake of essential minerals per day compared with the mopane caterpillar (*Imbrasia belina*)

Mineral	Intake recommendation for 25-year-old males (mg per day)*	Mopane caterpillar (mg per 100 g dry weight)
Potassium	4 700	1 032
Chloride	2 300	–
Sodium	1 500	1 024
Calcium	1 000	174
Phosphorus	700	543
Magnesium	400	160
Zinc	11	14
Iron	8	31
Manganese	2.3	3.95
Copper	0.9	0.91
Iodine	0.15	–
Selenium	0.055	–
Molybdenum	0.045	–

Note: * Dietary reference intakes (DRIs): recommended dietary allowances and adequate intakes, minerals, Food and Nutrition Board, Institute of Medicine, National Academies.

Source: Bukkens, 2005.

6.1.7 Vitamins

Vitamins essential for stimulating metabolic processes and enhancing immune system functions are present in most edible insects. Bukkens (2005) showed for a whole range of insects that thiamine (also known as vitamin B1, an essential vitamin that acts principally as a co-enzyme to metabolize carbohydrate into energy) ranged from 0.1 mg to 4 mg per 100 g of dry matter. Riboflavin (also known as vitamin B2, whose principle function is metabolism) ranged from 0.11 to 8.9 mg per 100 mg. By comparison, wholemeal bread provides 0.16 mg and 0.19 mg per 100 g of B1 and B2, respectively. Vitamin B12 occurs only in food of animal origin and is well represented in mealworm larvae, *Tenebrio molitor* (0.47 µg per 100 g) and house crickets, *Acheta domesticus* (5.4 µg per 100 g in adults and 8.7 µg per 100 g in nymphs). Nevertheless, many species have very low levels of vitamin B12, which is why more research is needed to identify edible insects rich in B vitamins (Bukkens, 2005; Finke, 2002).

Retinol and β-carotene (vitamin A) have been detected in some caterpillars, including *Imbrasia (=Nudaurelia) oyemensis*, *I. truncata* and *I. epimethea*; values ranged from 32 μg to 48 μg per 100 g and 6.8 μg to 8.2 μg per 100 g of dry matter for retinol and β-carotene, respectively. The levels of these vitamins were less than 20 μg per 100 g and less than 100 μg per 100 g in yellow mealworm larvae, superworms and house crickets (Finke, 2002; Bukkens, 2005; Oonincx and Poel, 2011). Generally, insects are not the best source of vitamin A (D. Oonincx, personal communication, 2012). Vitamin E featured in the palm weevil larvae, for example, which boasts 35 mg and 9 mg per 100 g of α-tocopherol and β+γ tocopherol, respectively; the daily recommended intake is 15 mg (Bukkens, 2005). The vitamin E content in ground and freeze-dried silkworm powder (*Bombyx mori*) is also relatively high, at 9.65 mg per 100 g (Tong, Yiu and Liu, 2011).

6.1.8 Fibre content

Insects contain significant amounts of fibre, as measured by crude fibre, acid detergent fibre and neutral detergent fibre. The most common form of fibre in insects is chitin, an insoluble fibre derived from the exoskeleton. A significant amount of data is available on the fibre content of insects, but it has been produced by various methods and is not easily comparable (H. Klunder, personal communication, 2012). Finke (2007) estimated the chitin content of insect species raised commercially as food for insectivores, and found it to range from 2.7 mg to 49.8 mg per kg (fresh) and from 11.6 mg to 137.2 mg per kg (dry matter).

Chitin, the main component of the exoskeleton of an insect, is a long-chain polymer of N-acetyl glucosamine – a derivative of glucose. Chitin is much like the polysaccharide cellulose found in plants, which is largely believed to be indigestible by humans, although chitinase has been found in human gastric juices (Paoletti *et al.*, 2007). Chitin has also been associated with defence against parasitic infections and some allergic conditions. The above study, carried out among Italians, showed an absence of chitinase activity in 20 percent of cases. Chitinase activity is more prevalent in tropical countries where insects are regularly consumed; there may be a lower rate of chitinase activity in Western countries due to the absence of chitin in the diet. Some argue that chitin acts like a dietetic fibre (Muzzarelli *et al.*, 2001), and this could imply a high-fibre content in edible insects, especially species with hard exoskeletons (Bukkens, 2005).

6.2 BEEF VERSUS INSECTS: AN EXAMPLE OF THE MEALWORM

Finke (2002) explored the nutritional value of several insect species, including the yellow mealworm (*Tenebrio molitor*). The larvae of the beetle have been mentioned as a promising option for mass rearing in Western countries because the species is endemic in temperate climates and easy to farm on a large scale, it has a short life cycle, and farming expertise is already available, particularly in the pet food industry. In the study by Finke (2002), insects were fasted for 24 hours to void their intestinal tract. The following conclusions were made (on a dry weight basis except for moisture and energy):

- **Macronutrient composition.** The fat content of beef is higher than that of mealworm larvae. Beef has slightly lower moisture content than mealworms and is marginally higher in protein and metabolizable energy.
- **Amino-acids.** Beef is higher in a.o. glutamic acid, lysine and methionine and lower in a.o. isoleucine, leucine, valine, tyrosine and alanine, compared with mealworms.
- **Fatty acids:** Beef contains more palmitoleic, palmitic and stearic acid than mealworms, but far higher values in essential linoleic acids were present in mealworms. Howard and Stanley-Samuelson (1990) analysed the phospholipid fatty acid composition of the adult *T. molitor* and found that over 80 percent of these fatty acids consisted of palmitic, stearic, oleic and linoleic acids. Finke (2002) found the same fatty acids in high amounts in *T. molitor* larvae. Polyunsaturated fatty acids are mostly found as phospholipids (Howard and Stanley-Samuelson, 1990).

- **Minerals.** Mealworms contain comparable values of copper, sodium, potassium, iron, zinc and selenium.
- **Vitamins.** Mealworms have generally higher vitamin content than beef, with the exception of vitamin B12.

TABLE 6.7
Average approximate analysis of selected *Tenebrio molitor* and beef as a percentage of dry matter except for moisture content

	T. molitor[1]	Beef
Moisture (% of fresh weight)	61.9	52.3
Protein	49.1	55.0
Fat	35.2	41.0
Metabolizable energy (kcal/kg)	2 056	2 820

Notes: [1] Mean body mass 0.13 g. Data presented based on a single analysis.
Source: Adapted from Finke, 2002, and USDA, 2012, by D. Oonincx.

TABLE 6.8
Average amino acid content of *Tenebrio molitor* and beef (amounts in g/kg dry matter unless stated otherwise)

Amino acid	*T. molitor* g/kg dry matter	Beef g/kg dry matter
Essential		
Isoleucine	24.7	16
Leucine	52.2	42
Lysine	26.8	45
Methionine	6.3	16
Phenylalanine	17.3	24
Threonine	20.2	25
Tryptophan	3.9	–
Valine	28.9	20
Semi-essential		
Arginine	25.5	33
Histidine	15.5	20
Methionine + cysteine	10.5	22
Tyrosine	36.0	22
Non-essential		
Alanine	40.4	30
Aspartic acid	40.0	52
Cysteine	4.2	5.9
Glycine	27.3	24
Glutamic acid	55.4	90
Proline	34.1	28
Serine	25.2	27
Taurine (mg/kg)	210	–

Source: Adapted from Finke, 2002, and USDA, 2012, by D. Oonincx.

The extent to which generalizations can be made about the nutrient content of *T. molitor*, presented in tables 6.7, 6.8 and 6.9, is limited, since data were from a single study and insect growth and development and nutritional composition depend on the specific diet of the insect (Davis and Sosulski, 1974; Anderson, 2000; Finke, 2002). *Tenebrio molitor* larvae, for example, need a dietary carbohydrate concentration of at

TABLE 6.9
Fatty acid content of *Tenebrio molitor* and beef on a dry matter basis

Fatty acid	Saturation	*T. molitor*[1]	Beef
Essential			
Linoleic	Omega-6 polyunsaturated	91.3	10.2
Linolenic	Omega-3 polyunsaturated	3.7	3.9
Arachidonic	Omega-6 polyunsaturated	–	0.63
Non-essential			
Capric	Saturated	–	1.05
Lauric	Saturated	< 0.5	1.05
Myristic	Saturated	7.6	13
Pentadecanoic	Saturated	< 0.5	–
Palmitic	Saturated	60.1	99
Palmitoleic	Omega-7 monounsaturated	9.2	17
Heptadecanoic	Saturated	< 0.5	–
Heptadecenoic	Omega-7 monounsaturated	0.8	–
Stearic	Saturated	10.2	48
Oleic	Omega-9 monounsaturated	141.5	159
Arachidic	Saturated	0.8	–
Eiconenoic	Omega-9 monounsaturated	–	0.63
Others		0.5	–

Notes: Hyphens indicate values that are not available. Values with inequalities indicate the detection limit of the assay; contents were lower than this limit. [1] Data based on a single analysis.

Source: Adapted from Finke, 2002, and USDA, 2012, by D. Oonincx.

least 40 percent to develop, and optimal growth is reached when the insect is grown on diets containing 70 percent carbohydrates (Behmer, 2006). Additionally, larvae grow and develop faster when a water source is available than when reared on dry food only (Urs and Hopkins, 1973a). Larvae reared in the presence of moisture are, moreover, heavier; this difference in weight is due not to higher water content but to a higher fat content because although insects can be fed on low-value organic waste streams it will affect their nutritional values, resulting in values lower than shown in tables 6.8 and 6.9.

6.3 INSECTS AS PART OF DIETS

6.3.1 The role of insects in food regimens: traditional diets

Traditional foods are those accepted by a community – through habit and tradition – to be desirable and appropriate sources of food. Traditional foods are accessible locally and within a given natural environment from farming or wild harvesting and constitute important elements in dietary regimens worldwide.

> The food systems of indigenous people show the important role of a diversified diet based on local plant and animal species and traditional food for health and well-being. In most cases, the increase of processed and commercial food items over time results in a decrease in the quality of the diet. Countries, communities or cultures that maintain their own traditional food systems are better able to conserve local food specialties with a corresponding diversity of crops and animal breeds. They are also more likely to show a lower prevalence of diet-related diseases (FAO, 2009b).

People in Africa, Asia and Latin America eat insects as regular parts of their diets. They may do so not only because conventional meats such as beef, fish and chicken are unavailable and insects therefore are vital sources of protein, but also because insects are considered important food items, often delicacies.

The problem is not simply convincing the West to consume insects, but also making sure that traditional practices of eating edible insects do not disappear as food regimens westernize. In countries where edible insects constitute regular elements in traditional diets, the shift towards Western foods constitutes a real threat to entomophagy. To counter this, efforts are being made to merge the traditional practice of insect eating with more popular foods. In Mexico, for example, it is not uncommon to find tortillas enriched with yellow mealworms, a traditional source of protein (Aguilar-Miranda *et al.*, 2002) (Box 6.5).

BOX 6.5

Don Bugito: creative and traditional Mexican food cart

Monica Martinez is a 36-year-old artist. Using art as a means, she wants to convince people to consider insects as a viable food source. This is the driving factor behind Don Bugito – launched in 2011 – a street food cart project which sells edible insect treats that are healthy for both people and the planet at street parties, festivals and food fairs. Inspired by prehispanic and contemporary Mexican cuisine, Don Bugito features a creative and traditional use of edible insects, grown organically and naturally in California, where Martinez is based. "San Francisco's foodie culture and its large Asian and Latino communities – whose cuisines already include edible insects – make the city a natural testing ground," says the artist. The cart features familiar Mexican ingredients – soft, blue corn tortillas, chilies and cheeses – along with protein-rich insects also found in prehispanic fare. The plump larvae of the wax moth fill tacos, along with peppers and a mint-cilantro salsa (Campbell, 2011). Martinez serves additional toasted crickets and, for dessert, caramelized mealworms on top of Mexican vanilla ice cream.

Source: Sweet, 2011.

6.3.2 How important are edible insects for protein intake in traditional diets?

The importance of edible insects on a global scale is difficult to estimate. Statistics and information are scarce and only available from a few, very specific studies. Nevertheless, such studies can provide an idea of how important edible insects are in various food systems and offer insights into the possibilities for developing the sector at a global scale.

Among indigenous peoples, insect gathering can be an important activity for food acquisition (see Chapter 3). A co-study of the Centre for Indigenous Peoples' Nutrition and Environment and FAO evaluated the nutritional and cultural importance of various traditional food items of 12 indigenous communities[11] from different parts of the world (see Table 6.11) (Kuhnlein, Erasmus and Spigelski, 2009). It found that the nutritional importance of insects in the Ingano community in Columbia was particularly significant. For example, the mojojoy larvae of May beetles and June beetles, which are both eaten in the Ingano community, are particularly rich in fat. Hormigas (Formicidae ants) also provide important sources of energy, and can be collected year round. The community described the attributes of the insects as follows:

- **Hormigas** (Formicidae ants). They are nutritious, very popular, improve growth, strengthen immune defences and provide proteins, vitamins and minerals.
- **Mojojoy** (May or June beetles). They are nutritious, improve growth and act as a medicine for pulmonary affections, their fat helps prevent pulmonary problems, and they provide proteins, vitamins and minerals.

[11] Ainu (Japan), Awajun (Peru), Baffin Inuit (Canada), Bhil (India), Dalit (India), Gwich'in (Canada), Igbo (Nigeria), Ingano (Colombia), Karen (Thailand), Maasai (Kenya), Nuxalk (Canada) and Pohnpei (Federated States of Micronesia).

Leaf-eating and litter-feeding invertebrates provide many Amerindian groups with important, underappreciated food sources. In the Amazon Basin, at least 32 Amerindian groups use terrestrial invertebrates as food (Paoletti *et al.*, 2000). The consumption of invertebrates provides significant amounts of animal protein (see Table 6.10), especially during lean times when fish and game are scarce. The Guajibo, for example, who live at the savannah border (at Alcabala Guajibo, Amazonas, Venezuela) rely mostly on insects, especially grasshoppers and larvae of the palm weevil *Rhynchophorus palmarum*. During the rainy season (July to August) over 60 percent of their animal protein is derived from insects. By selecting these small invertebrates, Amerindians choose their animal food from food webs in the rainforest that have the highest energy flow and which constitute the greatest renewable stock of readily available nutrients. The consumption of leaf-eating and litter-feeding invertebrates by forest-dwelling peoples as a means of acquiring protein, fat and vitamins offers a new perspective for the development of sustainable animal food production.

TABLE 6.10
Annual consumption of invertebrates in the Tukanoan village of Iapu (Rio Papuri, Vaupes, Columbia), composed of about 100 people

Name	Mean fresh weight consumed (kg/year)	Percentage of total number of invertebrates consumed
Atta soldiers and queens (three species)	100	29.3
Syntermes soldiers (three species)	133	39.0
Caterpillars (five species)	96	28.1
Vespidae larvae and pupae (three species)	2	0.60
Melaponinae larvae and pupae (one species)	1.5	0.44
Rhynchophorus palmarum larva	6	1.7
Beetle larvae boring on wood and dead wood (four species[1])	2.5	0.73

Note: [1] Four species (Scarabaeidae, Buprestidae, Cerambycidae, Passalidae).
Source: Paoletti *et al.*, 2000.

A mid-twentieth century study in the southwest of the Democratic Republic of the Congo found that animal protein was obtained from large game, crickets and grasshoppers during the dry season and largely from caterpillars during the rainy season (see also Chapter 2) (Adriaens, 1951). Fish, rodents, reptiles and various insect larvae were eaten year-round. The estimated production of dried caterpillars in the district of Kwango between 1954 and 1958 was nearly 300 tonnes per year. Moreover, in six provinces of the Democratic Republic of the Congo, insects constituted an average of 10 percent of the animal protein in daily diets (up to 15–22 percent in western provinces), with fish and game meat the two primary sources of protein, at 47 percent and 30 percent, respectively (Gomez, Halut and Collin, 1961). More recently, it was found that in the city of Kananga in the country's southwest, 28 percent of the inhabitants ate insects, mainly termites, caterpillars and beetle larvae, at an average of 2.4 kg of insects per month (Kitsa, 1989). Only palm beetle larvae and soldier termites (20 percent of the edible insect species), however, were available in markets throughout the year, while the remainder, in particular caterpillars and flying termites, were only seasonally available (December to April).

Food consumption data on wild, underused, indigenous and traditional plant and animal foods, however, remain limited and fragmented (FAO, 2010c). As the importance of food biodiversity becomes increasingly acknowledged, more research needs to be directed towards the consumption and composition of a wide variety of foods, including insects. Specifically, research is needed on the nutritional composition of wild, underused, indigenous and traditional insects and food biodiversity more generally, and data need to be compiled in accessible databases. To this end, an international network for data on food biodiversity was set up in 2010 within the INFOODS network (Box 6.1).

TABLE 6.11
Traditional food items of four indigenous communities from different parts of the world: the Awajun (Peru), the Ingano (Colombia), the Karen (Thailand) and the Igbo (Nigeria)

Insects as traditional foods	English name	Local name
Awajun, Peru		
Coleoptera	Palm weevil larvae	Bukin
Hymenoptera (*Brachygastra* spp.)	Wasp larvae	Ete téji
Hymenoptera (Formicidae)	Ant	Maya
Ingano, Colombia		
Hymenoptera: *Atta* spp.	Leaf-cutting ant	Hormiga arriera
Coleoptera	Beetle	Mojojoy
Karen, Thailand		
Orthoptera: Gryllidae (*Gryllus bimaculatus*)	Field cricket	Xer-lai-zu-wa
Igbo, Nigeria		
Coleoptera	Beetle	Ebe
Isoptera: Termitidae (2 spp.)	Termite	Aku-mkpu, aku-mbe
Coleoptera: Curculionoidea (3 spp.)	Palm weevil larvae (palm, raffia palm)	Akpa-nkwu, akpa-ngwo, nzam
Orthoptera: Gryllidae	Cricket	Abuzu
Orthoptera: Acrididae	Locust	Wewe, igurube

Source: Kuhnlein, Erasmus and Spigelski, 2009.

6.4 SUSTAINABLE DIETS

> Sustainable diets are those diets with low environmental impacts which contribute to food and nutrition security and to healthy life for present and future generations. Sustainable diets are protective and respectful of biodiversity and ecosystems, culturally accepted, accessible, economically fair and affordable; nutritional adequate, safe and healthy; while optimizing natural and human resources. (FAO, 2010b)

The need to feed a growing global population inevitably places continuous pressure on crop production, which in turn contributes further to the degradation of natural resources (FAO, 2009a). Difficulties arising from climate change, moreover, are set to compound present problems in production. Currently, FAO activities on sustainable diets explore linkages and synergies among food biodiversity, nutrition, food composition, food production, agriculture, urban agriculture (the Food for the Cities programme) and sustainability. The underlying objective is to improve food and nutritional security and provide more ecologically sound food recommendations to consumers and policymakers, including clarifying what is meant by an environmentally sustainable food system (FAO, 2009b). Edible insects as food fit comfortably within this environmentally sound scenario (see Chapter 5) and, by extension, ought to be considered prime candidates as both food staples and supplements, as well as more generally for their role in sustainable diets.

6.5 EDIBLE INSECTS IN EMERGENCY RELIEF PROGRAMMES

According to the UN's Standing Committee on Nutrition, the largest single contributor to disease is malnutrition. In emergency situations, disease can often contribute to or be a direct result of malnutrition. This not only involves sustaining the *quantity* of food people get, but also the *quality* of food. Not enough (or too much) food, the wrong types of food, and the body's response to a wide range of infections that result in the malabsorption of nutrients or the inability to use nutrients properly to maintain health are all factors influencing malnutrition. From a clinical point of view, malnutrition is characterized by the inadequate or excess intake of protein, energy and micronutrients such as vitamins. This definition also includes the frequent infections and disorders that are the result of an inadequate diet (WHO, 2013).

In areas where food insecurity is salient – 70 countries around the world – fortified blended food products (FBFs) are typically distributed to the most vulnerable peoples. FBFs are blends of partially precooked and milled cereals, such as soya, beans and pulses, fortified with micronutrients. Special adaptations may contain vegetable oil or milk powder. Corn soya blend is the main blended food distributed by the UN's World Food Programme, although wheat soya blend is also used. FBFs are largely designed to provide protein and micronutrient supplements in food assistance programmes. They are also commonly used in World Food Programme Supplementary Feeding and Mother and Child Health programmes (Pérez-Expósito and Klein, 2009).

The problem, however, is that the principal ingredients of FBFs (such as soy) are generally not part of traditional diets, nor, in many countries, are they locally available crops, making them ill-suited from nutritional, social and ecological points of view, particularly within the framework of sustainable diets (FAO, 2010b). Considering the protein and micronutrient content of many edible insects, their minimal ecological impact, their availability and, above all, their cultural appropriateness in a large majority of developing countries where food insecurity is a primary concern, their use in FBFs ought to be considered.

BOX 6.6

WinFood: alleviating childhood malnutrition by improved use of traditional foods

WinFood, a project funded by the Consultative Research Committee for Development Research and Danida in Denmark, aims to develop nutritionally improved foods for infants and young children, based on the improved use of traditional foods. Farmed vegetables, fruit and animal-source foods are nutritious but expensive, and consumption is limited in this target group. An unbalanced diet with too few non-staple foods, moreover, leads to the inadequate intake of iron, zinc and vitamin A in particular and is a major cause of childhood malnutrition. The idea of WinFood is to contribute to alleviating child malnutrition by focusing on traditional food systems based on semidomesticated or wild indigenous plant or animal foods (such as fruits, roots, small fish, snails, frogs and insects) and on traditional processing practices such as fermentation, germination and the soaking of staple and non-staple foods.

The WinFood concept is being developed through parallel studies in Cambodia and Kenya, two countries with very different cultural and ecological settings. Based on the results, generic guidelines for a WinFood strategy will be developed for implementation at the household level or through local small and medium-sized enterprises.

Edible insects – as traditional local foods in both Cambodia and Kenya – play an important role because they are locally available and are important sources of zinc and iron. For this reason, two WinFood products have been developed:

- WinFood Cambodia, which consists of rice, fish and spiders (*Haplopelma albostriatum*), among other food items;
- WinFood Kenya, which typically includes amaranth grain, maize, fish and termites (*Macrotermes subhyalinus*).

While results are promising, the lack of food standards for edible insects remains a major obstacle to further development (see chapters 10 and 14).

Source: N. Roos, personal communication, 2012.

Weaver ants

Cochineal on Opuntia cactus, La Palma, Canary Islands

FAO/YASUYOSHI CHIBA

A dense swarm of locusts as seen during spraying operations, Madagascar

ARNOLD VAN HUIS

Termite mound

JOOST VAN ITTERBEECK

Harvesting weaver ant larvae and pupae in Laos

ARNOLD VAN HUIS

Cricket trapping, Laos

PAUL VANTOMME

Seperating mealworms from chaff, the Netherlands

HARMKE KLUNDER

A woman harvests grasshoppers in Laos

LAUREN HEATON

Developing feed for aquaculture systems from Black solider fly larvae

MARCUS HARRISON

Shield bug snacks

FAO - PATRICK DURST

Bamboo borers cooked and prepared for sale at a local market in Chaing Mai, Thailand

BENJAMIN HARINK

Inside a Japanese ornamental bug store

DAWN STARIN

A variety of insects for sale as street food in Bangkok, Thailand

FRANK SHULZ

Chapulines vendors in Oaxaxa, Mexico

DAVID SKINNER

Antique chocolate covered ant tin, USA

FAO/GIULIO NAPOLITANO

Caterpillars for sale in Kinshasa, Democratic Republic of the Congo

AFTON HALLORAN

Insect snacks and candies for sale, Canada

MICHAEL FULLER

Scorpions for sale, China

AGRIPROTEIN

Feeding trials with magmeal made from the common housefly

JOSH EVANS

Bee larvae granola with bee larvae yoghurt, Nordic Food Lab, Copenhagen

7. Insects as animal feed

7.1 OVERVIEW

In 2011, combined world feed production was estimated at 870 million tonnes, with revenue from global commercial feed manufacturing generating approximately US$350 billion globally. FAO estimates that production will have to increase by 70 percent to be able to feed the world in 2050, with meat outputs (poultry, pork and beef) expected to double (IFIF, 2012). Despite this, little has been said about the opportunities insects offer as feed sources (Box 7.1). At present, ingredients for both animal and fish feed include fishmeal, fish oil, soybeans and several other grains.

A major constraint to further development are the prohibitive costs of feed, including meat meal, fishmeal and soybean meal, which represent 60–70 percent of production costs. Another problem is manure disposal, which is becoming a serious environmental problem; it is not uncommon for large amounts of manure to be stockpiled in open-air lots, swarming with flies.

BOX 7.1

International Feed Industry Federation and FAO: looking for new, safe proteins

The International Feed Industry Federation (IFIF) is a global organization mandated to play a coordinating role in promoting the sustainable supply of safe, healthy feed in the global feed industry. Its function is fundamentally important in developing countries, especially where national feed associations and sectors are weak or non-existent. In the late 1990s, the IFIF received Codex Alimentarius non-governmental organization (NGO) status, which was a first step towards improving the ways in which government regulates the industry. During this time, the IFIF began developing a close working relationship with FAO. Participation in Codex and FAO meetings has enabled the IFIF to follow up on the development and harmonization of international codes, standards and practices that affect feed manufacturers worldwide. Specifically, the IFIF has developed the *Codex Code of Practice of Good Animal Feeding*; participated in the Codex electronic working group on animal feeding; supported an FAO Expert Consultation on Protein Sources for the Animal Feed Industry; and developed the joint biannual Global Feed and Food Congress. Additionally, together with FAO, the IFIF has developed a *Manual of Good Practices for the Feed Industry* and established a meeting point for feed associations and feed regulators at an annual International Feed Regulators Meeting. The IFIF is convinced that sound scientific and technological advances will make a difference in ensuring that food is safe, plentiful and affordable to all.

Fishmeal prices are on the rise (Figure 7.1). Increased demand in 2010 and 2011 led to sharply higher prices and, although demand softened in late 2011 and early 2012, prices remain high. For small farmers, this means that fishmeal is less accessible. At the same time, aquaculture is the fastest-growing animal-food producing sector and will need to expand sustainably to keep up with increasing demand for fish.

At present, around 10 percent of global fish production goes to fishmeal (i.e. either whole fish or fish remains resulting from processing) and is used mainly in aquaculture (FAO, 2012b). South America is the biggest producer of fishmeal, through its catch of anchoveta. Anchoveta catch is highly variable because it is dependent on the El Niño climatic cycle. Production (catch) peaked at 12.5 million tonnes in 1994 but declined to 4.2 million tonnes in 2010 and is expected to fall further.

Insects have a similar market to fishmeal; they are employed as feed in aquaculture and livestock and also used in the pet industry. Recent high demand and consequent high prices for fishmeal, together with increasing production pressure on aquaculture, has led to research into the development of insect proteins for aquaculture and livestock (which could eventually supplement fishmeal). Meanwhile, aquaculture is growing and fishmeal is declining rapidly as a source of feed (Box 7.2) because of decreased supplies of industrially caught fish due to tighter quotas, additional controls on unregulated fishing, and greater use of more cost-effective dietary fishmeal substitutes (FAO, 2012b). The search for alternative and *sustainable* proteins is an issue of major importance that needs viable solutions in the short term, making insects an increasingly attractive feed option.

FIGURE 7.1
International wholesale market price for fish oil and fishmeal, CIF Hamburg

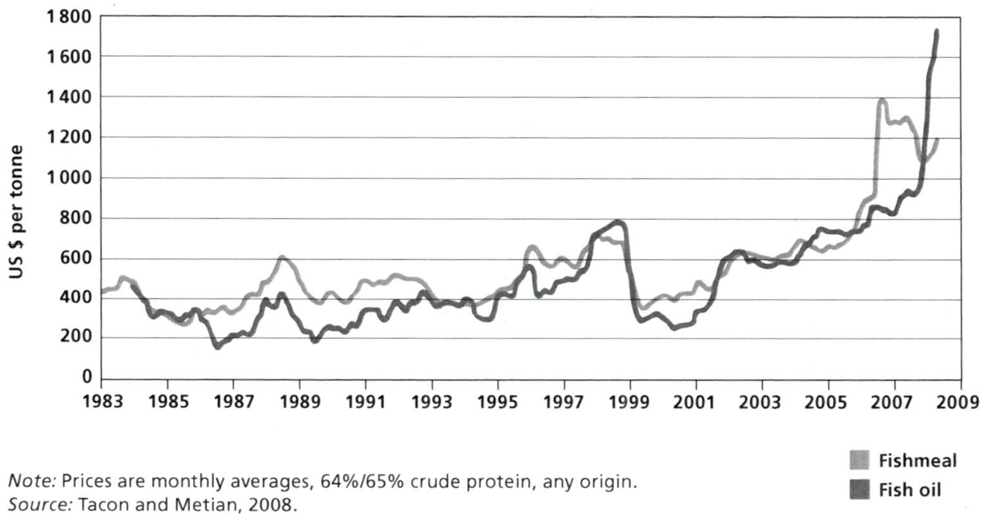

Note: Prices are monthly averages, 64%/65% crude protein, any origin.
Source: Tacon and Metian, 2008.

BOX 7.2
Fish for non-food uses

The amount of captured fish for non-food uses increased between 1976 and 1994. Since then it has declined, however, from 34 percent of the total catch in 1995 to about 26 percent in 2009, and as a consequence the total amount destined for fishmeal and fish oil has also declined (from about 30 percent to 20 percent). In 2008, aquaculture used 61 percent of world fishmeal production and 74 percent of fish-oil production. However, fishmeal use in aquafeeds has fallen from 19 percent in 2005 to 13 percent in 2008, and it has been predicted that it will decrease to 5 percent by 2020.

7.2 POULTRY AND FISH FED WITH INSECTS

Insects are natural food sources for many fish and poultry. Chickens, for example, can be found picking worms and larvae from the topsoil and litter where they walk. There is a reason, too, why maggots are used as fish bait in recreational fishing. Given insects' natural role as food for a number of farmed livestock species, it is worth reconsidering their role as feed for specific poultry and fish species (Box 7.3).

BOX 7.3
Which insects are currently used in animal feed?

FAO's Animal Feed Resources Information System (now called Feedipedia) provides information about the use of insects as animal and fish feed, including insects such as the desert locust (*Schistocerca gregaria*), common housefly maggots (*Musca domestica*) and domesticated silkworm (*Bombyx mori*). Information on source, processing, feeding guidelines, feeding experiments, feeding guidelines and nutrients characteristics are available under the category "animal products".

However, many other insect species may also be well suited to industrial-scale feed production, such as Coleoptera, which are presently raised by ornamental collectors (see Chapter 2).

7.2.1 Poultry

The poultry industry has expanded rapidly in developing countries in the last two decades. Grasshoppers, crickets, cockroaches, termites, lice, stink bugs, cicadas, aphids, scale insects, psyllids, beetles, caterpillars, flies, fleas, bees, wasps and ants have all been used as complementary food sources for poultry (Ravindran and Blair, 1993). In developing countries, animal and plant proteins supply the amino acids (e.g. lysine, methionine and cystine) in poultry feed. Animal-based, protein-rich feed ingredients are generally made up of imported fish and meat or blood meal, while plant-based resources include imported oil cakes and leguminous grains. Termites have reportedly been used as feed for chickens and guinea fowl in Togo and Burkina Faso (see section 2.3) (Iroko, 1982; Farina, Demey and Hardouin, 1991).

Chitin, a polysaccharide found in the exoskeleton of insects, may have a positive effect on the functioning of the immune system (see section 10.3). By feeding insects to chickens, the use of antibiotics in the poultry industry – which may lead to human infection with drug-resistant bacterial strains (Box 7.4) – may be diminished.

BOX 7.4
Chicken consumption leading to human infection with highly drug-resistant ESBL strains

In the Netherlands, patients suffering from serious urinary tract or bloodstream infections were evaluated. One in five of these patients were infected with ESBL (extended spectrum beta-lactamase) bacteria genetically identical to bacteria found in chicken. ESBL-containing strains of bacteria produce enzymes that bring about resistance to antibiotics like penicillin and the cephalosporins. Two bacteria – *Escherichia coli* and *Klebsiella pneumoniae* – most commonly produce ESBL enzymes. About 35 percent of the human isolates contained poultry-associated ESBL genes. The use of antibiotics is higher in the Dutch poultry industry than in any other European country; consequently, ESBL prevalence is correspondingly high. The study also revealed that nearly all (94 percent) chicken in Dutch supermarkets and at poultry farms are infected with ESBL bacteria, possibly due to the common use of antibiotics in their feed. Research is needed to ascertain whether feeding chickens with insects (containing chitin) will make the use of antibiotics superfluous by strengthening the immune system.

Source: van Hall *et al.*, 2011.

Ravindran and Blair (1993) cited the use of soldier flies (*Hermetia illucens*) grown on manure and housefly pupae (*Musca domestica*) as replacements for soya meal in poultry diets. Likewise, studies have shown how silkworm pupae – byproducts of silk manufacturing – can replace fishmeal entirely in the diets of layer chickens (i.e. in egg production) and supplement chicken diets (50 percent). Grasshoppers and Mormon crickets (*Anabrus simplex*) can also replace fishmeal and soy meal entirely.

In South Kivu, the Democratic Republic of the Congo, Munyuli Bin Mushambanyi and Balezi (2002) explored the possibility of replacing extremely expensive meat meal – a 20 percent feed ingredient in poultry farming – with flour derived from cockroaches (*Blatta orientalis*) and termites (*Kalotermes flavicollis*). Their study showed that the insect-derived flour could replace the meat meal ingredient when incorporated in the feed.Ramos Elorduy *et al.* (2002) conducted similar experiments with mealworms (*Tenebrio molitor*), rearing them on low-nutritive waste products and feeding them to broiler chickens. The mealworms were able to transform the low-nutritive waste products into a high-protein diet, making *T. molitor* a promising source of alternative protein, in particular as a replacement of soybean meal in poultry feed. Similar results were found in trials with *Anabrus simplex*, *Acheta domesticus*, *Bombyx mori*, *Alphitobius diaperinus*, *Tribolium castaneum* and termites (Ramos Elorduy *et al.*, 2002).

In India, the poultry industry is one of the fastest-growing agro-businesses, but the use of expensive maize as a feed ingredient is threatening the survival of farmers. Feeding poultry with sericulture waste, which until now has only been used for biogas production and composting, showed better conversion rates than those obtained through the use of conventional feed stock (Krishnan *et al.*, 2011).

7.2.2 Fish

Insects as sources of fish feed remain underappreciated in most parts of the world. In Uganda, a vast array of ingredients are used as fish feed, including vegetables, grass,

FIGURE 7.2
The proportional use of different types of feed by Ugandan fish farmers

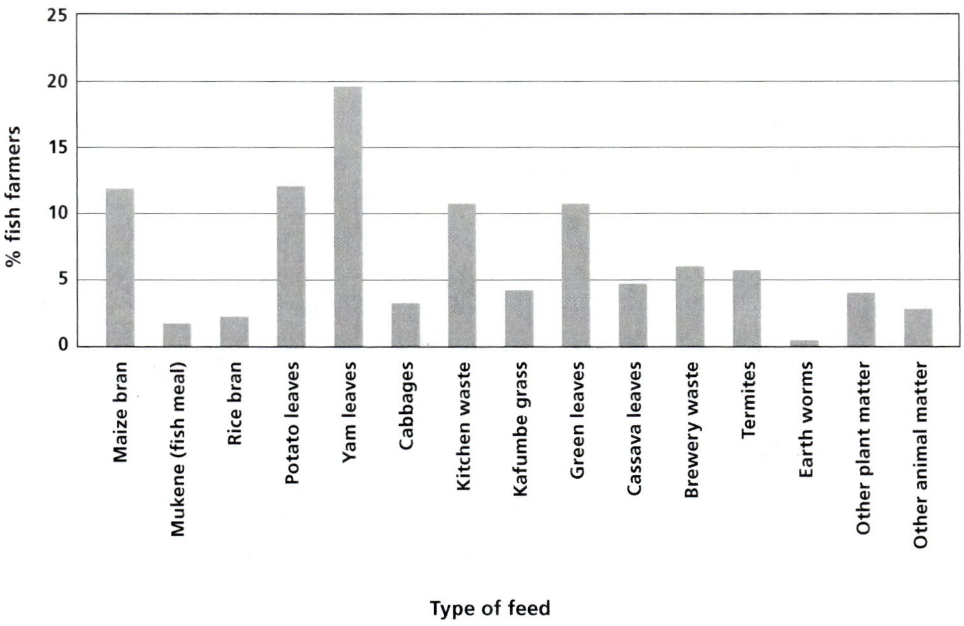

Source: Rutaisire, 2007.

cereals, cereal brans, oil seed cakes, industrial and kitchen wastes and fishmeal, as well as insects (Figure 7.2). The availability of most of these ingredients is seasonal (Rutaisire, 2007). Five percent of farmers use termites for feeding fish – either collecting the termites directly or purchasing them from collectors at a cost of US$0.27/kg – from March to April and from August to September. The quantity available depends largely on the number and size of termite hills on the farm, moonlight intensity and termite species. On average, a termite hill yields approximately 50 kg per year. In Southeast Asia it is very common to hang fluorescent lights above fish ponds. The light attracts the insects, which because of its reflection in the water, fall into the pond where they are eaten by fish. Wingless grasshoppers and crickets (which cannot float) are also used as fish bait, as are ant larvae and pupae (e.g. *Oecophylla smaragdina* in the Lao People's Democratic Republic) (J. Van Itterbeeck, personal communication, 2012).

7.3 KEY INSECT SPECIES USED AS FEED

Among the most promising species for industrial feed production are black soldier flies, common housefly larvae, silkworms and yellow mealworms. Grasshoppers and termites are also viable, but to a lesser extent. To date, these species are the most studied and account for the majority of the literature.

7.3.1 Black soldier flies

Black soldier flies (*Hermetia illucens*) (Diptera: Stratiomyidae) are found in abundance and naturally occur around the manure piles of large poultry, pigs and cattle. For this reason, they are known as latrine larvae. The larvae also occur in very dense populations on organic wastes such as coffee bean pulp, vegetables, distillers' waste and fish offal (fish processing byproducts). They can be used commercially to solve a number of environmental problems associated with manure and other organic waste, such as reducing manure mass, moisture content and offensive odours. At the same time they provide high-value feedstuff for cattle, pig, poultry and fish (Newton *et al.*, 2005). The adult black soldier fly, moreover, is not attracted to human habitats or foods and for that reason is not considered a nuisance. The high crude fat content of black soldier flies can be converted to biodiesel: 1 000 larvae growing on 1 kg of cattle manure, pig manure and chicken manure produce 36 g, 58 g and 91 g, respectively, of biodiesel (Li *et al.*, 2011). The possibility of recovering chitin after oil recovery is also being explored (see section 9.1).

Reducing populations of houseflies

Many environmental problems associated with manure storage and management can be solved by black soldier fly prepupae production. Sheppard *et al.* (1994) documented how the colonization of poultry and pig manure by black soldier flies can reduce populations of common housefly (*Musca domestica*) by 94–100 percent. Black soldier flies also make manure more liquid and thus less suitable for housefly larvae, and their presence is believed to inhibit oviposition by the housefly (Sheppard, 1983). While generally considered a nuisance, houseflies can also be reared as animal and fish feed.

Reducing manure contamination

Black soldier fly larvae are capable of converting residual manure proteins and other nutrients into more valuable biomass (e.g. animal feedstuff). In this way they reduce nutrient concentration and the bulk of manure residue. The harvested and processed black soldier fly larvae, valued at approximately US$200 per tonne, can also be more economically transported than manure (valued at US$10–20 per tonne) (Tomberlin and Sheppard, 2001). In confined bovine facilities, the larvae were found to reduce available phosporous by 61–70 percent and nitrogen by 30–50 percent (Sheppard, Newton and Burtle, 2008). In a field trial conducted in Georgia, United States, black soldier fly larval digestion of pig manure reduced nitrogen by 71 percent, phosphorous by 52 percent and potassium by 52 percent, and aluminium, boron, cadmium, calcium, chromium, copper, iron, lead, magnesium, manganese, molybdenum, nickel, sodium, sulphur and zinc by 38–93 percent. Thus, the larvae are able to reduce pollution potential by 50–60 percent or more. Foul odours produced by decomposing manure were also reduced or eliminated by black soldier fly larval digestion. This is because the species aerates and dries the manure, reducing odours. Additionally, the larvae modify the microflora of manure, potentially reducing harmful bacteria (Erickson *et al.*, 2004; Liu *et al.*, 2008). For example, larval activity significantly reduced *Escherichia coli* 0157:H7 and *Salmonella enterica* in hen manure (Erickson *et al.*, 2004). Sheppard, Newton and Burtle (2008) suggested that the larvae contain natural antibiotics similar to the larvae of the common green bottle fly (*Lucilia sericata*) used in maggot debridement therapy for cleansing human wounds, a method increasingly practised because of the prevalence of drug-resistant bacterial infections (Sherman and Wyle, 1996).

Black soldier flies as animal feed

The use of black soldier fly prepupae as animal feed should be seriously considered, not least for their reduced environmental footprint (Newton *et al.*, 1977; Sheppard *et al.*, 1994) (Box 7.5). Dried black soldier fly prepupae contain 42 percent protein and 35 percent fat (on a dry matter basis) (Newton *et al.*, 1977). Live prepupae consist of 44 percent dry matter and can easily be stored for long periods. As a component of a complete diet, they have been found to support good growth in chicks (Hale, 1973), pigs (Newton *et al.*, 1977) rainbow trout (*Oncorhynchus mykiss*) (St-Hilaire *et al.*, 2007), channel catfish (*Ictalurus punctatus*) (Pimentel *et al.*, 2004) and blue tilapia (*Oreochromis aureus*) (Sheppard *et al.*, 2008). In the case of rainbow trout, the larvae can replace 25 percent of fishmeal use and 38 percent of fish oil use. Instead of feeding insects to fish, insects can be reared on fish. Among the organic waste products, fish offal (entrails, etc.) can be fed to larvae. Compared with larvae fed on manure, lipid content increased by 30 percent and omega-3 fatty acids increased by 3 percent; both increases occurred within 24 hours (St-Hilaire *et al.*, 2007).

BOX 7.5

Increasing the sustainability of freshwater prawn production in Ohio

Freshwater prawn culture is increasingly popular in many temperate regions in the United States. Freshwater prawns have great potential for the diversification of Ohio farms. In the past ten years, interest in this product has increased due to rising demand for locally raised products, the growing desire among consumers to know where and how their food is produced, the uniqueness of the product, and increases in production rates for prawns based on new management and production practices.

Feed is the second highest contributor to variable production costs (the first being larval prawn procurement). Traditionally, most prawn farmers use a sinking catfish feed. With costs of these fishmeal-based diets continuing to rise, many animal nutritionists are looking for alternative protein sources to use in aquaculture feeds. One of these is black soldier fly larvae and their frass, or castings. For the first time in the United States, black soldier fly larvae are being cultured on a commercial scale in Yellow Springs, Ohio, by a company called EnviroFlight, where the first prawn diets using black soldier fly frass and wheat middlings as ingredients were produced.

The only notable difference was that the prawns fed the Enviroflight diet were slightly paler in appearance than those fed the traditional diet. Experienced prawn taste-testers

Continues

Box 7.5 continued

detected no flavour difference between the two products. Using a locally produced aquaculture feed has many benefits for Ohio freshwater prawn producers, as well as potential benefits for aquaculture producers growing other fish species. First, the cost of the feed is lower than currently commercially available alternatives. This will contribute to the economic efficiency of operations, especially considering that the cost of fishmeal is predicted to continue to rise. Given that the feed is produced in Ohio, fewer "food transportation miles" can be attributed to its production and distribution. Additionally, feeding prawns a diet devoid of fishmeal may open additional marketing opportunities for farmers, as some customers are opposed to the use of fishmeal in aquaculture feeds. Finally, given that the black soldier fly larvae eat dried distiller's grains, the production of this product actually aids the efficient reuse of waste/coproducts derived from another Ohio industry. This recycling of nutrients adds to the overall sustainability of the project.

Source: Tiu, 2012.

7.3.2 Common housefly larvae

Maggots – the larvae of the common housefly (*Musca domestica*) – develop predominantly in tropical environments. Maggots are important sources of animal proteins for poultry: they have a dry matter of 30 percent of their total wet larval mass, 54 percent of which is crude protein. Maggots can be offered fresh, but for intensive farming they are more convenient as a dry product in terms of storage and transport. Studies have shown that maggot meal could replace fishmeal in the production of broiler chickens (Téguia *et al.*, 2002; Hwangbo *et al.*, 2009). At the same time, maggot production can contribute to alleviating manure accumulation.

In rural Africa, maggots are natural food items for scavenging poultry. In Nigeria, for example, maggot production could provide an excellent source of animal protein for local poultry farms. Maggots are already fed live to chickens in Togo (Ekoue and Hadzi, 2000) and Cameroon (Téguia, Mpoame and Okourou, 2002). In South Korea, Hwangbo *et al.* (2009) explored the contribution of maggots to the meat quality and growth performance of broiler chickens and found that feeding diets containing 10–15 percent maggots can improve the carcass quality and growth performance of broiler chickens. In Nigeria, Awonyi, Adetuyi and Akinyosoye (2004) evaluated the replacement of fishmeal with maggot meal and found that diets in which 25 percent of fishmeal was replaced with maggot meal were most efficient in terms of average weekly weight gain and protein efficiency rate. At nine weeks, the live, dressed and eviscerated weights of the chickens, as well as the relative length, breadth and weights of the pectoral and gastrocnemius muscles, were not significantly affected by replacement with maggot meal. It was concluded that maggot meal is an inexpensive partial substitute for fishmeal in broiler-chick feeding.

The inclusion of maggot meal in livestock diets, however, raises concerns because common knowledge suggests that, in its adult form, *Musca domestica* is widely involved in the transmission of disease. The larvae develop in manure and decaying filth; for this reason, maggot meal in livestock diets raises bacteriological and mycological concerns. In Nigeria, Awoniyi, Adetuyi and Akinyosoye (2004) investigated fresh and nine-month-old stored samples of dried, milled housefly larvae for the presence of microbes to determine their suitability for inclusion in livestock diets. Their main conclusion was that stored maggot meal is prone to deterioration by fungi and bacteria if the moisture content is too high (in their study 23 percent, while the limit was 12 percent). They recommended drying to 4–5 percent moisture to minimize bacterial activity. After processing, protection from moisture absorption can be achieved by waterproof bagging (with cellophane or nylon) and heat-sealing.

7.3.3 Termites

Termites caught in the wild can be used to catch fish and birds. Silow (1983) reported from Zambia the use of snouted termites (*Trinervitermes* spp.) as fish bait in conical reed traps and as bait to attract insectivorous birds (such as guinea fowl, francolins, quails and thrushes). The birds were caught by setting a snare across the broken top of a termite mound, where soldiers mass for hours. However, rearing termites is very difficult and should not be recommended, also bearing in mind their high emissions of methane (Hackstein and Stumm, 1994).

7.3.4 Silkworms

In most developing countries, animal production is hindered by scarcity and the expense of fishmeal as a feed ingredient. Although sericulture produces vast amounts of pupae, research dealing with silkworm caterpillar meal as a feed ingredient is scanty. In Nigeria, Ijaiya and Eko (2009) analysed the possibility of substituting fishmeal (by 25, 50, 75 and 100 percent) with silkworm (*Anaphe panda*) caterpillar meal in relation to the growth, carcass haematology and economics of broiler chicken production and found that the growth performance of chickens was not affected by the incorporation of silkworm caterpillar meal. There were no significant differences in performance in terms of feed intake, body weight gain, feed conversion efficiency or protein efficiency ratio between dietary treatments. Silkworm caterpillar meal proved less expensive than conventional fishmeal, making it well suited in economic terms as a substitute.

7.3.5 Mealworms

Mealworms (such as *Tenebrio molitor*) are already raised on an industrial scale. They can be grown on low-nutritive waste products and fed to broiler chickens. Ramos Elorduy *et al.* (2002) reared *T. molitor* larvae on several dried waste materials of different origins. They used three levels of larvae (0, 5 and 10 percent dry weight) in a 19 percent protein content sorghum–soybean meal basal diet to evaluate feed intake, weight gain and feed efficiency. After 15 days there were no significant differences between treatments. Mealworms are promising alternatives to conventional protein sources, particularly soybean meal.

7.3.6 Grasshoppers in India

In India, research has been conducted on the use of grasshoppers as feed for farm animals. This is because conventional feed accounts for 60 percent of the total cost of raising farm animals, and also because there is a shortage of feedstuffs such as maize and soybean as a result of competition between humans and livestock for these resources. In addition, harvesting these food acridids in cropland and grasslands may allow a reduction in the use of harmful pesticides for their control. Four species of acridids were studied for their nutritional content: *Oxya fuscovittata*, *Acrida exaltata*, *Hieroglyphus banian* and *Spathosternum prasiniferum prasiniferum* (Anand, Ganguly and Haldar, 2008). The study found acridids to have a higher protein content compared with the conventional soybean and fishmeal available locally.

Rearing and mass production

The use of acridids as animal feed requires a huge biomass, which can only be obtained by mass rearing in insect farms. Das, Ganguly and Haldar (2009) studied the space required for mass rearing *Oxya fuscovittata* and *Spathosternum prasiniferum prasiniferum*. The use of jars with a volume of 2 500 cm^3 and a density of 10 000 insects per m^3 for *O. fuscovittata* and 7 100 insects per m^3 for *S. pr. prasiniferum* resulted in mortality rates of 12 percent and 15 percent, respectively. The smaller size of *S. pr. prasiniferum* meant that more could be kept per unit area compared with *O. fuscovittata*. Das, Ganguly and Haldar (2010) also determined the optimum temperature and photoperiod to mass-rear

Oxya hyla hyla and experimented with the use of grasshopper manure for soil fertility enhancement. They found that the percentages of nitrogen, phosphorous and potassium were similar for acridid species compared with those for commonly used animal manure.

Feeding trials with fish and poultry

Feeding trials on certain fish species revealed that diets in which 25 percent and 50 percent of fishmeal was replaced with acridid meal produced results as good as the control diet comprising 100 percent fishmeal. All growth parameters measured for the selected fish were higher for the formulated feed containing acridid meal than for those fed with market-available diets. This indicates that acridids could prove a successful meal replacement for conventional fishmeal.

Japanese quail (*Cotornix japonica japonica*) were fed with various diets in which *Oxya* meal gradually replaced fishmeal. For a range of growth parameters, the best results were obtained with the diet in which 50 percent of fishmeal was replaced with *Oxya* meal. Moreover, fecundity (i.e. the number of eggs laid per female) was significantly higher compared with the control treatment.

Thus, among the selected acridids, two nutritionally rich species of the genus *Oxya* (*O. fuscovittata* and *O. hyla hyla*) have the ability to produce substantial biomass due to their elevated rates of fecundity and fertility. It is estimated that *Oxya* could replace at least 50 percent of fishmeal to feed fish and poultry birds. These results support the idea of establishing acridid farms in which *O. fuscovittata* and *O. hyla hyla* are mass-reared using *Sorghum halepense* grasses and *Brachiaria mutica* plants as food. The transition to acridid tissues would be relatively simple, ensuring the provision of a constant source of feed for developers to supplement the diets of livestock intended for human and non-human consumption. Moreover, if acridids are popularized as alternative food and feed sources, this could significantly lower the rate of overexploitation of fishmeal and consequently decrease the demand/supply ratio of fishmeal, helping to reduce market prices (Haldar, 2012).

8. Farming insects

8.1 DEFINITIONS AND CONCEPTS

Defined broadly, agriculture includes farming both animals (animal husbandry) and plants (agronomy, horticulture and forestry in part) (FAO, 1997b). The concept of insect farming, however, is relatively new in development circles, including FAO. Insects are reared in a designated area (i.e. a farm) and the insects' living conditions, diet and food quality are controlled. Farmed insects are kept in captivity, or "ranched", and are thus isolated from their natural populations. The term semi-cultivation, as it applies to insects, is defined in section 4.4.

The words *rearing* and *breeding* are often confused. The word breeding is more often used in livestock production than in entomology. Strictly speaking, *rearing* refers to tending the animals, while breeding refers to their reproduction. *Breeding* often refers to producing better offspring: that is, genetically improving the stock by selecting specimens in a population with certain desired characteristics. But keeping insects under confined conditions can also have a genetic effect on populations through inbreeding depression, founder effect, genetic drift and laboratory adaptation, such that they often no longer much resemble wild populations.

The distinction between livestock and minilivestock is not always clear: *minilivestock* implies small animals raised for domestic use or profit (not as pets), especially on a farm. These can be small mammals, amphibians, reptiles or invertebrates, including insects (Paoletti, 2005). According to Hardouin (1995), "these animal species include both vertebrates and invertebrates, which can be terrestrial or aquatic by nature, but of a weight usually under 20 kg, and that these animals must have a potential benefit, either nutritionally or economically". Conversely, *livestock* generally refers to cattle, poultry, sheep, llamas, alpacas, goats, camels, horses and other similar animals raised for domestic use or profit, but not as pets.

8.2 INSECT FARMING

Most edible insects are harvested in the wild, but a few insect species have been domesticated because of their commercially valuable products. Silkworms and bees are the best-known examples. Sericulture – the practice of rearing silkworms for the production of raw silk – has its origins in China and dates back 5 000 years. The domesticated form has increased cocoon size, growth rate and efficiency of digestion, and is accustomed to living in crowded conditions. The adult can no longer fly and the species is completely dependent on humans for survival. Both bee larvae and silkworm pupae are eaten as byproducts (Box 8.1). Additionally, some insect species are reared for the pet-food industry. For example, mealworms and crickets are reared primarily as pet food in Europe, North America and parts of Asia.

BOX 8.1

Dual production systems (fibre and food): the example of the silkworm

Colombia. In sericulture, the pupae of the domesticated silkworm (*Bombyx mori*) are considered byproducts and are good sources of food for both humans and animals. With

Continues

Box 8.1 continued

an estimated annual production of 1.2–1.4 million silkworm cocoons per ha of mulberry bushes and one pupa weighing 0.33 g (dry weight), the average yield of pupal byproduct is 400–460 kg per ha (DeFoliart, 1989). In addition, the frass (a waste that insects pass after digesting plant parts) can be used as fertilizer or as feed for pond fish.

India. Seriwaste is only used for biogas production and composting. Researchers are experimenting with feeding seriwaste to poultry (Krishnan *et al.*, 2011). The poultry industry is one of the fastest-growing agro-businesses in India, yet sustainable feed products with high conversion rates are not widely available. Krishnan *et al.* (2011) argued that seriwaste is extremely viable because silk waste is not toxic and has even better conversion rates than conventional feedstocks.

Kenya. A project in Kenya successfully linked forest conservation and livelihood improvement (Raina *et al.*, 2009). By commercializing insects such as the mulberry silkworm, local forest communities were able to sell the silk produced, which proved to be a valuable alternative source of cash income. Leftover pupae were fed to chickens. These benefits gave local communities incentives to better manage their surrounding forest habitat.

Madagascar. A local NGO, the Madagascar Organization of Silk Workers (SEPALI), along with its United States partner, Conservation through Poverty Alleviation, is implementing a programme to alleviate local pressure on the newly established Makira Protected Area by aiding local farmers in the production of artisanal silk from endemic moths. In 2013, SEPALI will develop the Pupae for Protein project. People in the Makira area eat some of the types of reared silkworms and pupae. Once farmers produce 4 000 pupae and select 200 for further rearing, the remaining 3 800 can be boiled, sautéed, dried or ground into a calcium-rich protein powder. In fact, 3 800 pupae are approximately equal in weight to one red-ruffed lemur, an endangered species.

Another commercially viable product obtained from farmed insects is carminic acid. Carminic acid derives from the cochineal insect (*Dactylopius coccus*), which is domesticated on the cactus *Opuntia ficus-indica* var. Atlixco (see section 2.4). The acid is used as red dye in human food and in the pharmaceutical and cosmetic industries.

Insects are also reared in agriculture to either combat insect pests or for pollination. In biological control, large rearing companies mass-produce beneficial insects such as predators and parasitoids (Box 8.2). These insects are often sold to fruit, vegetable and flower farmers to combat insect pests and are also used in large estate crops, for example egg parasitoids (*Trichogramma* spp.) and larval parasitoids (*Cotesia flavipes*) to combat sugarcane borers. Bumblebees (*Bombus* spp.) and honeybees (*Apis* spp.) are reared worldwide to help farmers pollinate crops and fruit orchards.

BOX 8.2

Biological control and natural pollination

The large-scale production of natural enemies to combat agricultural pests and bees to pollinate crops is a global business. Koppert Biological Systems is the international market leader in the field of biological crop protection and natural pollination. The company develops and markets pollination systems (bees and bumblebees) and IPM programmes for protecting high-value crops.

Natural enemies of insect pests, also known as biological control agents, include predators, parasitoids and pathogens. Predators (e.g. ladybirds) feed on their prey, resulting in the death of the organism. A parasitoid develops inside the host insect

Continues

Box 8.2 continued

and eventually kills it (e.g. parasitic wasps). Predators and parasitoids can be mass-reared and then released in the field or in glasshouses to contain agricultural pests, thereby minimizing damage to crops. Insects are also reared for the *in vivo* production of pathogenic nematodes and viruses. These insect products are excellent examples of non-chemical, non-toxic, non-hazardous and environmental friendly plant protection measures. They are commonly included in IPM strategies applied to control major insect pests in a large number of food and fibre crops.

A common method used in IPM is the sterile insect technique, in which large numbers of sterile insects are released into the environment, competing with wild males for female insects. When a female mates with a sterile male it will bear no offspring, thus reducing the next generation's population. Repeated release of sterile insects can eradicate or contain a population. The technique has been used successfully to eradicate the screw-worm fly (*Cochliomyia hominivorax*), a livestock pest in areas of North America, and to control the medfly (*Ceratitis capitata*) in Central America, which has caused extensive damage to a range of fruit crops.

In temperate regions there are companies that produce large numbers of insects as pet food and fish bait. The species most used are crickets (*Gryllodus sigillatus*, *Gryllus bimaculatus* and *Acheta domesticus*), mealworms (*Zophobas morio*, *Alphitobius diaperinus* and *Tenebrio molitor*), locusts (*Locusta migratoria*), sun beetles (*Pachnoda marginata peregrine*), wax moths (*Galleria mellonella*), cockroaches (*Blaptica dubia*) and maggots of the housefly (*Musca domestica*). Some companies even produce Mighty Mealys™, or giant mealworms, which are *T. molitor* larvae treated with juvenile hormones. The hormone suppresses pupation and allows the larvae to grow to a size of about 4 cm, making them ideal as pet food and bait.

In addition, some insects serve medicinal purposes. The common green bottlefly (*Lucilla sericata*), for example, is produced for use in maggot therapy. Live, disinfected fly larvae are introduced into soft tissue wound(s) of humans or animals to clean out the necrotic tissue (debridement) and disinfect the traumatized area. House dust mites are also produced commercially for allergy-testing. The Research Institute of Resource Insects of the Chinese Academy of Forestry in Kunming has conducted extensive research on the rearing of insects for medical applications (Feng *et al.*, 2009), and the use of insects as food in space is also being examined (Box 8.3).

Other reasons for rearing insects include research into plant breeding and chemical control (e.g. screening of pesticides and testing side-effects on non-target arthropod species). Insects are also produced for educational and recreational purposes, for example in zoos and butterfly gardens. In some countries, insects serve as pets – such as walking sticks and singing or fighting crickets in Chinese culture (see section 2.1), and scarab beetles, such as stag beetles (Lucanidae) and rhinoceros beetles (Dynastinae) in Japan, Thailand and Viet Nam.

The potential uses of insects are vast. Recently, the use of insects for the bioconversion of manure and waste has been explored (see section 7.4). It would be useful to engage industries already producing insects, for example as pet food, to promote production for animal feed and human consumption (e.g. mealworm, locusts and crickets).

8.3 INSECT FARMING FOR HUMAN CONSUMPTION

8.3.1 Tropics

The best example of rearing insects for human consumption in the tropics is cricket farming. In Thailand, two species are produced: the native cricket (*Gryllus bimaculatus*) and the house cricket (*Acheta domesticus*). The native cricket is interesting from an economic standpoint, but the taste and quality of the house cricket is generally thought to be superior (Y. Hanboonsong, personal communication, 2012).

The methods used in cricket farming in the Lao People's Democratic Republic, Thailand and Viet Nam – veteran producers of crickets – are very similar. In these countries, crickets are reared simply in sheds in one's backyard, and there is no need for expensive materials. In the Lao People's Democratic Republic and Thailand, concrete rings approximately 0.5 m in height and 0.8 m in diameter are used as rearing units, while plastic bowls are used in Viet Nam. In each "arena", a layer of rice hull (or rice waste) is placed on the bottom. Chicken feed or other pet food, vegetable scraps from pumpkins and morning glory flowers, rice and grass are used for nourishment. Plastic bottles are used to provide water; a plate of water filled with stones can also be used, with the stones preventing the crickets from drowning. Sticky tape or plastic tablecloths are stuck on the inner side of the walls just below the edge to keep the crickets from crawling out of the arena. Cardboard egg cartons, tree leaves and hollow logs are also used to create a larger amount of space for the crickets. Females lay their eggs in small bowls filled with sand and burned rice husks. After a time, these bowls are moved to another receptacle, where a new generation is reared. Each bowl is covered with a layer of rice hull to maintain a suitable incubating temperature. Crickets are kept from escaping by covering the arena, for example with mosquito netting; this also prevents other animals, such as geckos, from entering. The rearing areas are surrounded by a "moat" – a narrow strip of water containing very small fish – that prevents ants from entering (Yhoung-Aree and Viwatpanich, 2005; J. Van Itterbeeck, personal communication, 2008).

8.3.2 Temperate zones

In temperate zones, insect farming is largely performed by family-run enterprises that rear insects such as mealworms, crickets and grasshoppers in large quantities for pet food. Because the species are frequently reared in close, confined spaces, climate control is often applied, as high temperatures may cause the desiccation of soft-bodied larvae.

The rearing of high quantities of insects, either for consumption as whole insects and/or as protein extracts, is possible in industrialized countries. Critical elements for successful rearing include greater knowledge of biology, rearing conditions and artificial diet formulation (Wang *et al.*, 2004, Feng and Chen, 2009; Schneider, 2009). Diets can be altered to increase nutritional value (Anderson, 2000) and adapting the light regime can optimize production; for example, exposing crickets to 24 hours of light per day can increase cricket production (Collavo *et al.*, 2005). Such issues warrant further research.

BOX 8.3

Insect proteins in space

It has been suggested that insects could be used as a protein source in space flights. Scientists in China, Japan and the United States are looking seriously into this food resource for space travel and use in space stations. China is planning to build a terrestrial model of a bio-regenerative life support system that makes use of silkworms (DeFoliart, 1989; Katayama *et al.*, 2008; Hu, Bartsev and Liu, 2010). Species like *Agrius convolvuli, Stegobium paniceum* and *Macrotermes subhyalinus* have also been proposed (Katayama *et al.*, 2005).

Cohen (2001) criticized the lack of professional appreciation for insect rearing and proposed formalizing insect rearing and insect food science and technology as academic disciplines. High-quality rearing is essential for the widespread use of insects as human food.

Another major challenge is rearing insects in large numbers, which requires the development of automation processes. These are being explored for silkworms

(Ohura, 2003). Robert Kok and colleagues at McGill University, Canada, are conducting research on the optimal design of insect farms for large-scale production (Kok, 1983; Kok, Shivhare and Lomaliza, 1990) (see Table 8.1). Nevertheless, major impediments to full-scale rearing include cost and the still uncertain nature of the supply of waste streams and the ability to guarantee a consistent, high-quality product.

TABLE 8.1
Favourable characteristics of insects for automated production systems

Social structure of populations	Reaction to humans
Gregarious	Readily habituated
Small territories	Little disturbed
Males affiliated with female groups	Non-antagonistic, no disagreeable odour
Intra and interspecies agonistic behaviour	**Parental behaviour**
Non-agonistic to con-specifics	Egg guarding
Non-agonistic to non-specifics	Precocial young
Altruistic	Young easily separated from adults
Sexual behaviour	**Ontogeny**
Male initiated	Short developmental cycle
Sex signals via movement or posture	High survival of immatures
Pheromonally induced	High oviposition rate
Promiscuous	High potential of biomass increase/day
Easy to propagate	Low vulnerability to diseases/parasites
Feeding behaviour	**Locomotory activity and habitat choice**
Generalist feeder	Non-migratory
Feeds on common items	Sessile or small home range
Non-cannibalistic	Limited agility
Accepts artificial diet	Wide environmental tolerance
Endogenous feeding satiation	Ecological versatility

Source: Kok, 1983; Gon and Price, 1984.

8.4 INSECT FARMING FOR FEED

Insects are much more efficient in converting feed to body weight than conventional livestock and are particularly valuable because they can be reared on organic waste streams (e.g. animal slurries). Research into rearing insects as food and feed on a large scale remains a priority. Current production systems are still too expensive. A study in the Netherlands (Meuwissen, 2011) suggested that the production of mealworms is still 4.8 times as expensive as normal chicken feed. In particular, labour and housing costs for large-scale feed production facilities are much higher for insects than for the production of chicken feed.

8.5 RECOMMENDATIONS ON INSECT FARMING

The Expert Consultation Meeting on Assessing the Potential of Insects as Food and Feed in Assuring Food Security, held at FAO headquarters in Rome in January 2012, made recommendations for rearing insects, including suggestions on species and strain collection; household production; training in insect farming; the choice, cost and reliability of feedstock; safety, health and environmental issues; and strategic issues for industrial-scale insect farmers.

8.5.1 Species and strain collection for food and feed

The Expert Consultation Meeting agreed that rearing practices in tropical countries should employ local species because they pose virtually no risk to the environment, there is no need for climate control, and such local species are likely to be more culturally accepted. Selection criteria should involve ease of rearing, taste, colour and whether they can be used as feed. In temperate zones, cosmopolitan species like the house cricket (*Acheta domesticus*) should be used, or those that do not pose environmental risks.

Industrial-scale production was defined in the meeting as a minimum reach of 1 tonne per day of fresh-weight insects. Species destined for mass production, moreover, should possess certain characteristics, including a high intrinsic rate of increase; a short development cycle; high survival of immatures and high oviposition rate; a high potential of biomass increase per day (i.e. weight gain per day); a high conversion rate (kg biomass gain per kg feedstock); the ability to live in high densities (kg biomass per m^2); and low vulnerability to disease (high resistance). Good candidates were considered to be the black soldier fly (*Hermetia illuscens*) for feed and the yellow mealworm (*Tenebrio molitor*) for both food and feed. Because of the vulnerability of production systems, heavy reliance on a single species is discouraged (Box 8.4). Finally, it was recommended to preserve parental genetic lines in case of culture crashes.

Other questions identified and discussed at the Expert Consultation Meeting included:

- Is the species amenable to large-scale automation (thereby reducing labour costs)?
- Can the species be contained in non-native areas? What are the biodiversity consequences if they are introduced to non-native areas?
- Is there a possibility of genetically improving species by selective breeding to obtain high-quality strains?
- What is the ecological footprint of the insects (e.g. GHG production)?
- What are the water requirements?

8.5.2 Household production

In the tropics, the emphasis should be on maximizing the productivity of traditional management systems. Procedures for small-scale farming should be developed – such as kits for home-use – so that people can easily start small-scale rearing facilities. Feed for insects should be obtained locally. The possibility of using available organic waste (or rest streams), for example, should be evaluated.

BOX 8.4

Difficulties in rearing crickets in the Netherlands

The insect-rearing company Kreca used to sell more than 10 000 boxes of crickets (*Acheta domesticus*) each week. In 2000, 50 percent of the crickets reared by the company died within 8–12 hours, a population crash never previously experienced. A densovirus was suspected to be the cause of cricket mortality, and a very thorough sanitation regimen followed. All diseased crickets were removed, the entire rearing facility was cleaned, and strict hygiene measures were imposed. In addition to the sanitation programme, the location of rearing was moved and the cricket eggs were washed thoroughly. These efforts, however, proved futile. As a consequence, cricket rearing was terminated. Heavy reliance on a single species is strongly discouraged, for many of the same reasons that monocultures should be avoided in agriculture – their high vulnerability to diseases and pests. Kreca now rears three cricket species: *A. domesticus*, *Gryllus bimaculatus* and *Gryllodus sigillatus*, with the latter showing the most economic promise.

8.5.3 Training in insect farming

Farmers can learn from each other's experiences. Cooperatives can be effective for information-sharing and should be promoted in both tropical and temperate countries. Workshops for sharing knowledge and to increase networking should also be organized.

In tropical countries, moreover, training could be conducted using the "farmer field school" approach, which has proven successful in other agricultural development settings and requires the involvement of local extension services. Insect farming should be incorporated into formal educational systems, including elementary and secondary schools and universities, with the aim of making people aware that insects can be farmed just like other livestock.

The Faculty of Agriculture at the National University in the Lao People's Democratic Republic already teaches cricket farming to its students (see Chapter 12). In Thailand, the Faculty of Agriculture, Khon Kaen University, has an undergraduate teaching course in industrial entomology, including edible insect farming. In addition, cricket farming, processing and marketing is taught in an annual international training course on using indigenous food resources for food security.

8.5.4 Choice, cost and reliability of feedstock

In choosing a feedstock it is important to know whether the insects are destined for feed or food. For insects to be used as feed, different (organic waste) side streams need to be evaluated. Insects intended for human consumption need to be fed feed grade or even food grade food if the insects are not to be degutted. Waste streams might not be a viable option for human consumption; this area demands further research. Finally, the feedstock needs to be inexpensive, locally available, of consistent quality and supply, and above all free of pesticides and antibiotics.

8.5.5 Safety, health and environmental issues

In food production, safety is paramount. The mistakes made in the livestock industry (e.g. the overuse of antibiotics) should serve as a lesson for insect rearers. Disease management strategies need to be preventive in nature. Human hazards related to production should be circumvented, such as passive vectoring of pathogens and the development of allergies among personnel in production units. The rearing system design should also minimize sensitivity to disease. Risk guidelines as well as sanitary standards need to be developed and implemented for each species.

8.5.6 Strategic issues for industrial-scale insect farmers

The success of the industry will hinge on its ability to set up a reliable and consistent production chain and, above all, on its ability to produce high-quality feed and food with high nutritional value. The following developments are recommended:

- creating an international society of producers of insects as food and feed to complement the existing Association of Insect Rearers for Biocontrol, with the possibility of a society publication;
- developing a code of practice/standards (possibly modelled on those of the mushroom industry) and product quality metrics to garner credibility;
- adopting a common language in the industry to assist communication with the general public;
- developing a marketing strategy that establishes which industries and consumers to target;
- creating a list of species that are "society approved" for use as human food;
- centralizing information, literature, methods and practices;
- liaising with relevant policymakers and researchers.

9. Processing edible insects for food and feed

9.1 DIFFERENT TYPES OF CONSUMABLE PRODUCT

After being wild-harvested or reared in a domesticated setting, insects are killed by freeze-drying, sun-drying or boiling. They can be processed and consumed in three ways: as whole insects; in ground or paste form; and as an extract of protein, fat or chitin for fortifying food and feed products. Insects are also fried live and consumed.

In countries where edible insects are traditionally eaten, food habits have shifted towards Western diets. To counter this, initiatives have been undertaken, for example, in Mexico, where tortillas are being enriched with yellow mealworm (Aguilar-Miranda *et al.*, 2002). This section gives examples of innovative projects that have developed promising edible insect products.

9.1.1 Whole insects

In tropical countries, insects are often consumed whole, but some insects, such as grasshoppers and locusts, require the removal of body parts (e.g. wings and legs). Depending on the dish, fresh insects can be further processed by roasting, frying or boiling. In the Lao People's Democratic Republic, among other countries, insects can be found in markets as ready-to-eat snacks or fried with lime leaves.

9.1.2 Granular or paste form

Grinding or milling is a common method for processing a large variety of foods. Soybeans, for example, are often transformed into tofu or other meat analogues. Meat is processed into products such as hamburgers and hot dogs, and fish into popular foods such as fish fingers. In much the same way, edible insects can also be processed into more palatable forms. They are often ground into paste or powder and added to otherwise low-protein foods to increase their nutritional value. An easy way to obtain powder is by drying and grinding the insects. In Thailand and the Lao People's Democratic Republic, chilli paste with crushed and ground giant waterbugs (*Lethocerus indicus*) is very popular as a main ingredient (and is known locally as jaew maeng da in the Lao People's Democratic Republic and nam phik in Thailand). The flavour of giant water bug is now reproduced artificially and is readily available. In societies where consumers are not accustomed to eating whole insects, granular or paste forms may be better accepted.

9.1.3 Extracted insect proteins

Western consumers may be reluctant to accept insects as a legitimate protein source because insects have never played a substantial role in their food culture. Extracting insect proteins for human food products – a process already being carried out – could be a useful way of increasing acceptability among wary consumers. In some cases, isolating and extracting insect protein is desirable to increase the protein content of a food product. However, supplementing food products with insects in such a way requires extensive knowledge of the properties of the extracted proteins. These properties include, among others, amino acid profile, thermal stability, solubility, gelling, foaming and emulsifying capacity. Separating extracted protein groups based on their solubility in solvents produces water-soluble and water-insoluble fractions, which can be used for specific applications in both the food and feed industries. Alternative methods are enzymatic processes to obtain proteins of specific chain lengths. Alternative methods for protein separation are fluidized bed chromatography and ultrafiltration.

At present, the cost of protein extraction is prohibitive. More research is required to further develop the process and to render it profitable and applicable for industry use. Wageningen University is conducting a programme on the sustainable production of insect proteins for human consumption (in the period 2010–2013) to further explore the possibility of extracting proteins from insects to fortify human foods. Under the project, dubbed Supro2, edible insects are reared on organic side streams, after which their proteins are separated, purified and characterized in order to tailor them for specific food products. Extracted insect proteins could also be considered in feed products, although the economic feasibility would need to be established.

9.1.4 Examples of promising edible insect products for human consumption

SOR-Mite (protein-enriched sorghum porridge)

The "developing solutions for developing countries" competition, organized by the Institute of Food Technologists, promotes the application of food science and technology and the development of new products and processes with the aim of improving the quality of life of people in developing countries. The first prize of the competition, awarded during the Annual Food Expo in Anaheim, United States, in June 2009, went to the SOR-Mite project, a sorghum mixture enriched with termites. The nutritionally weak grain, frequently consumed in many African countries, is low in proteins and fats and lacks several essential amino acids, such as lysine. For this reason, fortifying the grain with highly nutritious flying termites (*Macrotermes* species), easily gathered at the start of the rainy season, makes sense. The fermented mixture can be consumed as porridge at breakfast, lunch or dinner, depending on local preferences. Both raw materials are easily obtained locally (Institute of Food Technologists, 2011).

Termite crackers and muffins in Kenya

In the Lake Victoria region in East Africa, edible insects such as termites (Isoptera: Termitidae) and lake flies (Diptera, Chaoboridae, Chironomidae and Ephemeroptera) are abundant and provide important nutrition for both humans and livestock. Although their use is limited by their seasonal availability and high perishability, processing them with conventional cooking methods could extend their shelf life considerably and contribute to promoting entomophagy throughout the region. In a recent study carried out in the cross-border ecosystem, locally available insects were roasted, sun-dried, ground and mixed with other ingredients and processed into food products. Termite-based and lakefly-based crackers, muffins, meatloaf and sausages were found to have especially high potential for commercialization (Ayieko, Oriamo and Nyambuga, 2010).

Buqadilla

Buqadilla is an innovative snack under development for the Dutch market. It is a spicy Mexican leguminoceous food product made of chickpeas and lesser mealworms (40 percent). It was well received in several restaurants and canteens, where the product was tested, for its taste and smooth structure. The sustainable, healthy and exotic snack is an example of an accessible and culturally acceptable way for Western consumers to experience and appreciate edible insects as food (van Huis, van Gurp and Dicke, 2012).

Crikizz

Crikizz is another example of European products made with insects. Developed by Ynsect and French students, Crikizz are spicy, popped snacks based on mealworms and cassava. The mealworm composition varies from 10 to 20 percent in accordance with the product line ("classic" to "extreme"). According to focus groups, the taste is very pleasing and differs from other snacks, while the texture is as crunchy as other snacks. The prototype was made without preservatives or taste enhancers, and the high fat composition of mealworms removes the need for added fat. Crikizz won a prize in the national French contest Eco-trophélia 2012 for culinary innovation.

Processed mealworms for pet food, animal feed and human food

HaoCheng Mealworm Inc. in China specializes in the farming and sale of mealworms, superworms and maggots. The farm, established in 2002, consists of 15 rearing facilities and produces 50 tonnes of living mealworms and superworms per month. HaoCheng exports 200 tonnes of dried mealworms to Australia, Europe, North America and Southeast Asia each year.

The mealworms, superworms and maggots are sold live, dried, canned and in powder form. They have elevated protein content and can be used as additives for food as well as feed:

- Food. Mealworm powder can be worked into bread, flour, instant noodles, pastries, biscuits, candy and condiments. The insects can also be consumed whole as meals and side dishes, or processed into medicinal supplements to fortify the human body's immune system.
- Feed. Entire insects can be used as direct feeds and feed supplements for pets such as birds, dogs, cats, frogs, turtles, shrimps, scorpions, chilopods, ants, goldfish and wild animals (Hao Cheng Mealworm Inc., 2012).

9.1.5 Extracted fat

The removal of fat (and ash) in the production of insect products, such as insect meal, reduces the "stickiness" of the protein concentrate and prevents fatty acids (mostly unsaturated) from exposure to undesirable oxidation processes. The extracted fat can then be used for other purposes. Traditionally, the fat (e.g. oil) of some insect species is used extensively for frying meat and other food products (Box 9.1).

BOX 9.1

Termites: processing techniques in East and West Africa

- Winged termites are often fried in their own fat. Fried termites contain 32–38 percent protein (Tihon, 1946; Santos Oliveira *et al.*, 1976; Nkouka, 1987).
- In Uganda, termites are steamed in banana leaves.
- Termites are boiled or roasted after swarming and then sun-dried or smoke-dried, or both, depending on the weather (Silow, 1983).
- Sometimes termites are crushed with a pestle and mortar and eaten with honey (Ogutu, 1986). The fat residues of fried termites can be used to cook meat (Bequaert, 1921), a time-honoured practice among Azande and pygmies in the Democratic Republic of the Congo (Bergier, 1941).
- Pygmies put the oil derived from frying or pressing dried termites into tubes and use it to treat their body and hair (Costermans, 1955).
- In many East African towns and villages, sun-dried termites can be bought at local markets when in season (Osmaston, 1951; Owen, 1973).
- In Botswana, San women harvest winged termites (*Hodotermes mossambicus*), roasting them in hot ash and sand (Nonaka, 1996).

9.2 INDUSTRIAL-SCALE PROCESSING

There is a wealth of traditional and cultural knowledge on the uses of edible insects as food in tropical countries, yet production is largely concentrated in household and small-scale operations. In temperate countries, processing technology is virtually non-existent because edible insects are not recognized food and feed sources. If insects are to become a useful and profitable raw material in the food and feed industries, large quantities of quality insects will need to be produced on a continuous basis. This requires the automation of both farming and processing methods, which remains a challenge for the development of the sector (see Table 9.1).

TABLE 9.1
Important aspects of large-scale production of edible insects

Standardization	Processing methods monitored equally
Legislation	Regulations and guidelines developed for producers. This includes feed sourcing and its standards (no waste in the EU) and as other aspects, such as welfare, biosecurity in terms of escapees, disease management etc.
Shelf life	Final product easy to store and preferably having a long shelf life
Transportation	Final product easy to transport
Tradeoffs between quality and safety	During processing (nutritional) quality maintained or increased
Costs	Product compatible to alternatives on the current market

9.2.1 Examples of industrial insect farming

AgriProtein (South Africa) and Enviroflight (United States) are examples of companies developing industrial-scale insect farming.

AgriProtein

The prospect of farms processing insects for feed might soon become a global reality, due to a growing demand for sustainable feed sources. AgriProtein is leading a new industry called nutrient recycling, which uses organic waste to create protein that will help meet the increasing demand for animal feed. It is a global project focused on fish and meat farming to cater to the growing world population. By using common housefly larvae fed on abundant waste nutrient sources, AgriProtein has developed and tested a new large-scale and potentially sustainable source of protein. The bioconversion process takes low-cost waste materials and generates a valuable commodity.

The production process starts with rearing stock flies in sterile cages, each holding over 750 000 flies. Various types of waste are used, including human waste (faeces), abattoir blood and spent food. Depending on the species, a single female fly can lay up to 1 000 eggs over a seven-day period, which then hatch into larvae. Housefly larvae go through three life stages in a 72-hour period and are harvested just before becoming pupae. The harvested larvae are dried on a fluidized bed dryer, milled into flake form and packed according to customer preferences.

The product contains nine essential amino acids, with high levels of cystine and similar levels of lysine, methonine, threonine and tryptophane – similar to marine fishmeal. Potential big users would need vast quantities of the product – some pet food businesses alone could use over 1 000 tonnes per month.

The company started making only small amounts in the laboratory, but in recent years production has risen to hundreds of kg per day and will shortly exceed 1 tonne per daily production run. The end target is 100 tonnes of larvae per day. A first large factory would require an investment of US$8 million and the countries planned for roll-out are Germany, South Africa, the United Kingdom of Great Britain and Northern Ireland and the United States (see figures 9.1 and 9.2) (Agriprotein, 2012).

FIGURE 9.1
Agriprotein fly protein production process

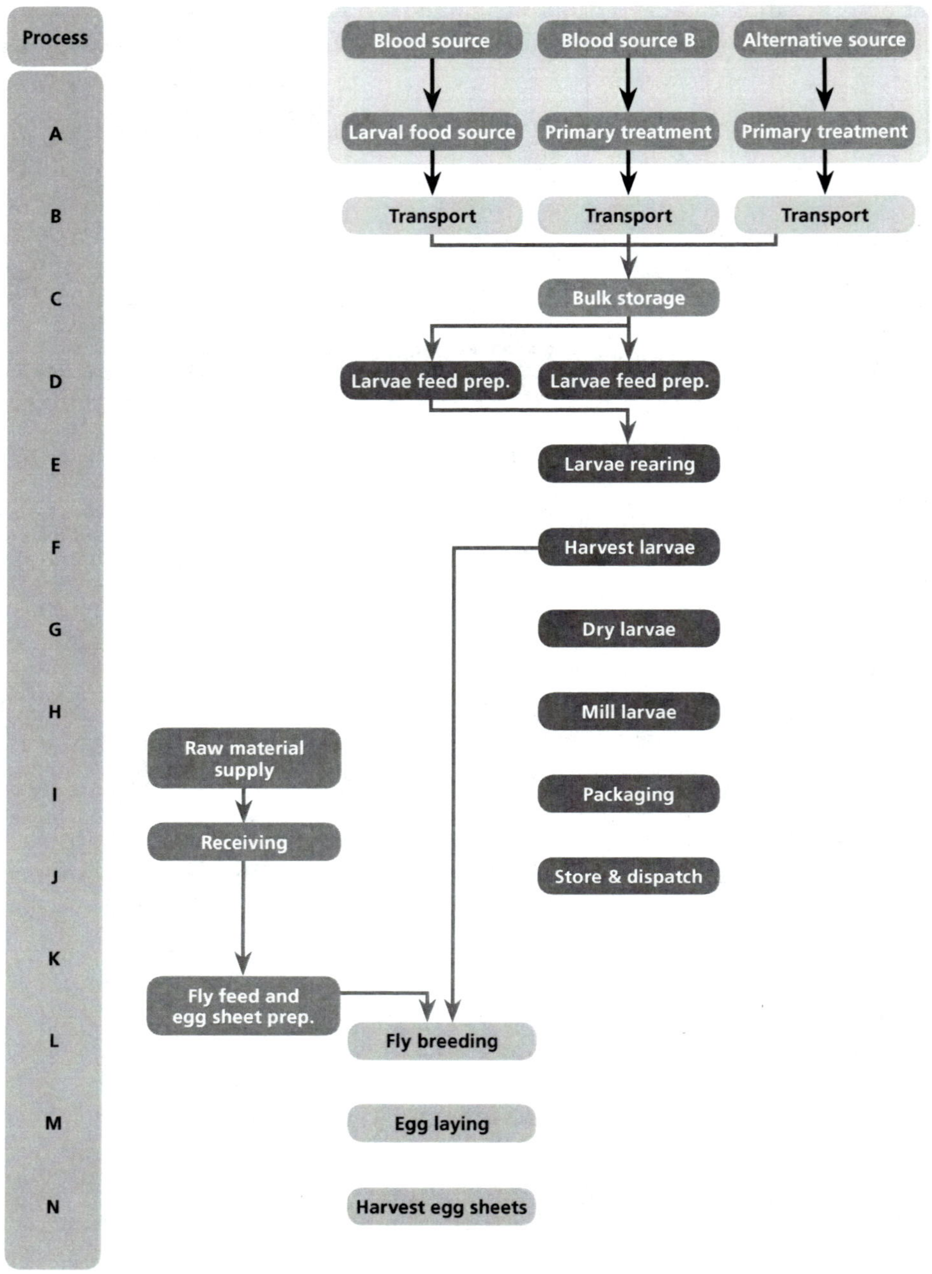

Based on: Agriprotein, 2012.

Enviroflight

Enviroflight is another producer of insects for animal feed. Enviroflight's objective is to produce animal and vegetable protein for aquaculture feeds.

Enviroflight uses dried distiller grains with solubles from ethanol plants and spent brewer grains. Black soldier flies (*Hermetia illucens*) feed off and bioconvert the grain. By doing so, the frass becomes a high-protein, low-fat feedstuff for tilapia, freshwater prawns, catfish and other omnivorous species. The material is also beneficial as a protein source for pigs and cattle.

FIGURE 9.2
Agriprotein value/production chain

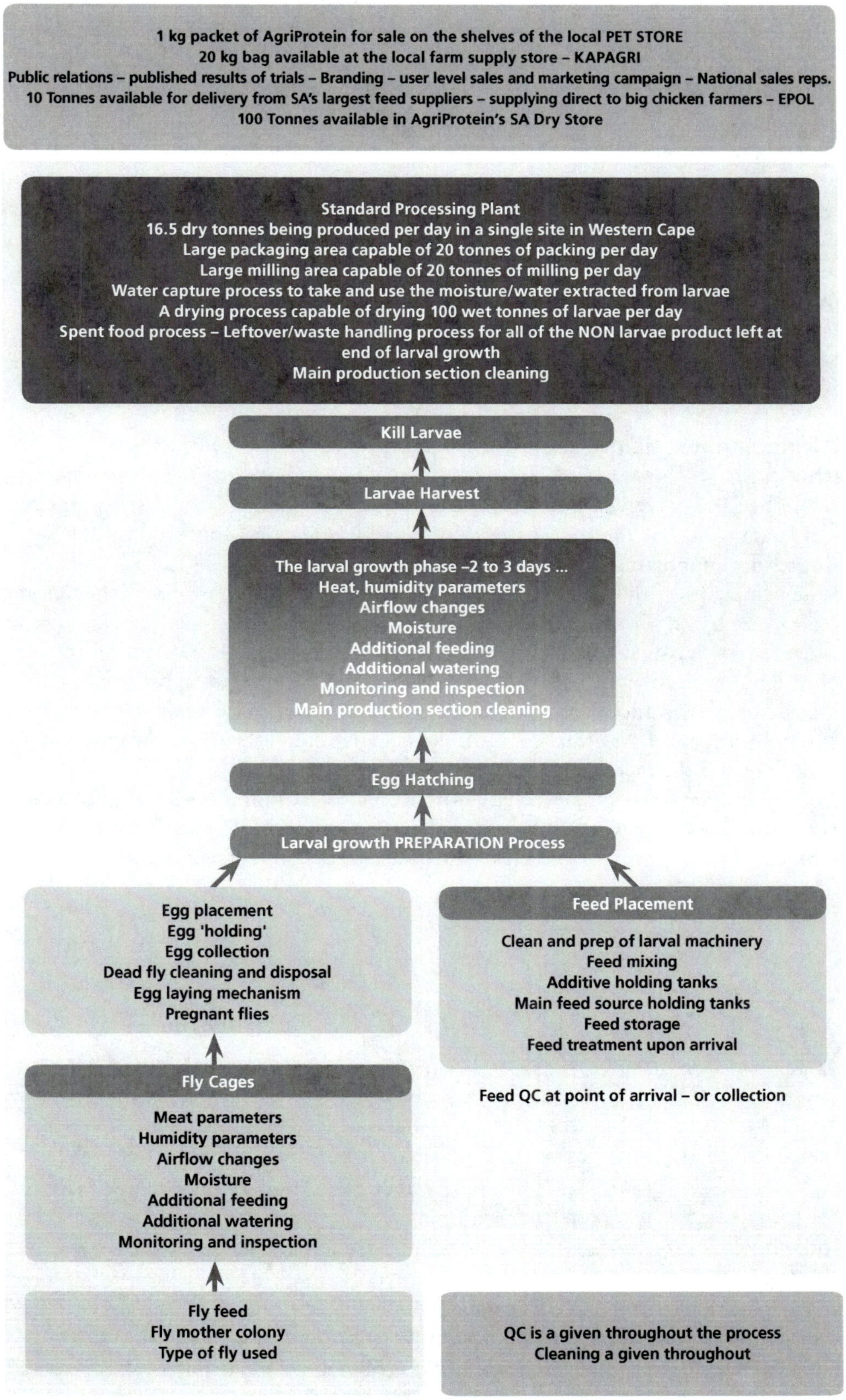

Source: Agriprotein, 2012.

The larvae are used as a high-protein, high-fat ingredient for carnivorous fish such as rainbow trout, perch, bass and bluegill. They are cooked, dried and converted into a meal that is 42 percent protein and 36 percent fat. The oils can be extracted, which boosts the protein content to above 60 percent. Enviroflight has developed and tested a number of feed formulations using insect meal and other locally available ingredients for complete diets for multiple species of fish.

A key to the process is that it prevents the creation of ammonia in the frass by stabilizing the material immediately after the insect larvae consume the brewers' or distillers' grains. This keeps the nitrogen fixed and eliminates odour; it also mitigates the formation of molds and mycotoxins. The frass is also beneficial as a natural, animal-safe fertilizer. The nitrogen, phosphorous and potassium levels are 5 percent, 3 percent and 2 percent, respectively, which is very good for vegetable growing.

The system comprises a proprietary bioreactor system and a breeding chamber. It is designed to be located anywhere in the world and can be optimized for operation in developing countries. The breeding chambers are able to produce mating events and eggs in any climate. Because of this there is a continual source of eggs in all weather conditions (G. Courtright, personal communication, 2012).

9.2.2 Industrial-scale processing for insects as food and feed in the Netherlands

The Netherlands is developing an innovative supply chain that includes large-scale insect farming and marketing the insect-derived products for food and feed. Research institutes are supporting this development process.

The principles of the circular economy and theories on environmental economics (Box 9.2) are based on an interrelationship between the environment, economics and the future scarcity of sufficient, nutritious and healthy food. The design of the insect supply chain is circular (Box 9.3). It is based on farming insects on organic waste and using the insects as a food or feed ingredient. This takes place against a background of growing demand for animal protein, the negative side-effects of conventional meat production, and the increasing problem of waste disposal.

Supply-chain partners, knowledge institutes, NGOs and national and regional governmental bodies have a roadmap for creating a prosperous insect industry by 2020 (see Figure 9.3). The aim for 2020 is to introduce farmed insects as ingredients for feed and food.

BOX 9.2

Environmental economics

Balasubramanian (1984) stated that "no longer is economics merely a science of production and distribution, it has to take into account the ecological repercussions of economic activities that could affect both production and distribution". Thus, economics should not merely be the study of how goods and services are produced; it should take into consideration the impacts of the use of resources on the environment. Any study on the economic content of production, distribution and development cannot be complete without touching on issues such as externalities, pollution, damage, exhaustion and depletion, among others. Environmental economics can therefore be defined as that "part of economics which deals with interrelationships between environment and economic development and studies the ways and means by which the former is not impaired nor the latter impeded" (Sankar, 2001). It is thus a branch of economics that discusses the impacts of interactions between humans and nature and finds human solutions to maintain harmony. Insects could play an important role in finding such solutions.

BOX 9.3

Application of edible insects: insects as the missing link in designing a circular economy[12]

To satisfy the growing demand for sufficient affordable and sustainable proteins, the following innovation processes are proposed:

- **Insects as bioconverter.** The feed, food and pharmaceutical industries use ingredients from insects grown on organic waste. There are possibilities for processing insects for applications in medicine, cosmetics, alcohol, etc.
- **Alternative designs for a viable and sustainable agricultural sector.** The agricultural sector is under pressure to obtain higher yields with less input. More and more companies are being forced to cease production because they cannot cope with this competitive battle, while scaling up is not an option. Insects farmed on organic waste can be an attractive and viable alternative for entrepreneurs.
- **Opportunity for innovative modern entrepreneurs.** The Netherlands is a global leader in the field of life sciences and has many entrepreneurs with relevant knowledge. These entrepreneurs can play a leading role in developing the sector.

The potential for marketing insect-derived products depends on the following conditions:

- reliable high volumes of production;
- fair and competitive market prices;
- resolutions to hurdles in legislation;
- permission to use (organic) waste or byproducts from the food and agricultural industry.

The major challenges are:

- **Scaling up.** Large-scale production units will be necessary to reduce costs. Insect products are currently substantially more expensive than regular meat products. Mealworms are approximately three times more expensive than pork and about five times more expensive than chicken. The main factor is labour and how to increase the competitiveness of alternative animal protein products. The use of insects in the feed sector in Europe costs about €100 per 100 kg. Attention must be paid to increased automation and mechanization processes in order to arrive at processed end products (i.e. insect flour instead of whole insects).
- **Increased market and consumer acceptance.** Critical factors are the development of market concepts and business cases for feed and food in close cooperation with partners in the supply chain. Market pull is needed to create a higher turnover for insect farms and enable them to invest in scaling their facilities.
- **Legislative national and international frameworks.** Legal frameworks need to allow the use of insects as ingredients for food and feed. The use of organic waste as a feed ingredient for insects is another legislative concern. Approval by the European Union (EU) Novel Food Regulation is an expensive and time-consuming procedure. In the short term, legislation does not allow for the use of protein extracts originating from insects. The Transmissible Spongiform Encephalopathies regulation at the European level also blocks the introduction of insects as an ingredient for feed.
- **Developing new cooperative and funding models.** The establishment of large-scale production facilities, pre-competitive collaboration and innovative approaches to market acceptance, as well as new forms of financing, are needed for the establishment of this new sector.

---

[12] Box contributed by Marian Peters of Venik.

FIGURE 9.3
Insects as the missing link: ecology designs a circular economy

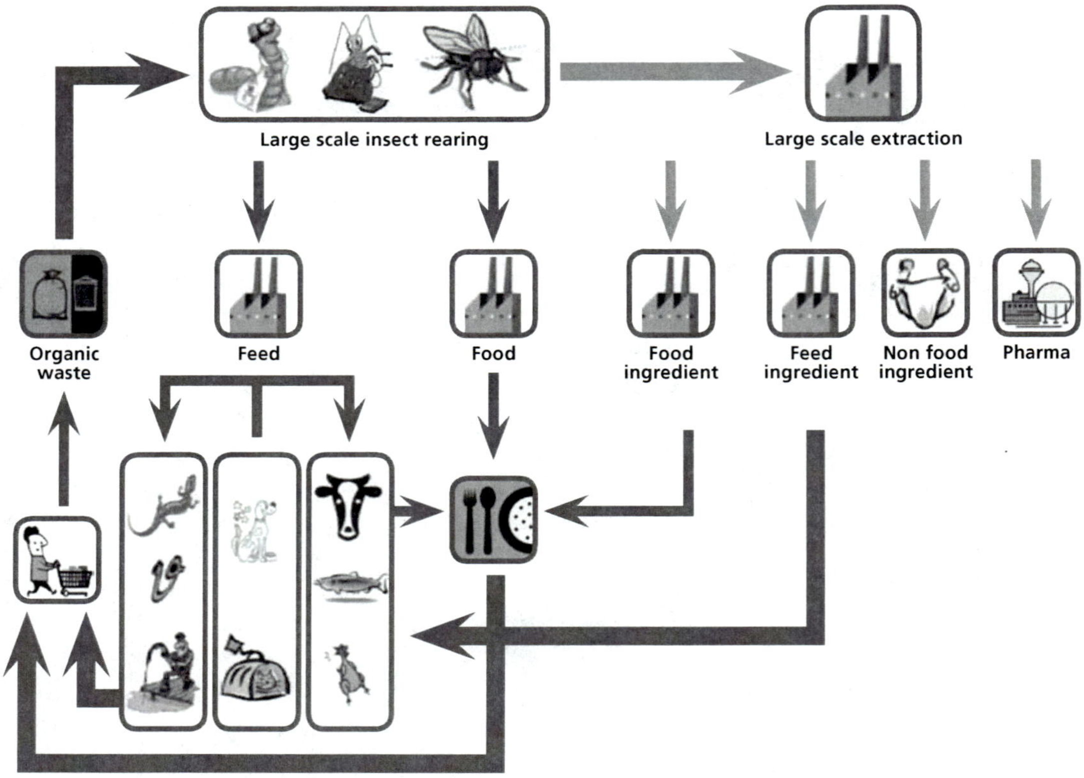

Source: M. Peters, personal communication, 2012.

10. Food safety and preservation

Food safety, processing and preservation are closely related. Insects, like many meat products, are rich in nutrients and moisture, providing a favourable environment for microbial survival and growth (Klunder *et al.*, 2012). Traditional processing methods, such as boiling, roasting and frying, are often applied to improve the taste and palatability of edible insects and have the added advantage of ensuring a safe food product. Cultural preferences and organoleptic (sensory) aspects play important roles in chosen preservation methods. Although a wide variety of modern preservation methods is available, specific measures to ensure a high-quality and safe food item may be required for different insect species, depending on their biological makeup. Determining optimal preservation methods will be a critical factor in the commercialization of edible insects on a global scale, be it for food or feed. In this chapter the main focus is on food; however, the same also applies for feed.

The Hazard Analysis Critical Control Points (HACCP) system, a science-based and systematic tool, identifies specific hazards and establishes control systems to ensure the safety of food (FAO/WHO, 2001a). Its focus is preventative in nature, rather than relying mainly on end-product testing. HACCP is recognized worldwide as a system for quality assurance, identifying, evaluating and controlling physical, chemical and biological hazards throughout the production process. The system can be applied across the food chain, from primary production to final consumption.

As well as increasing food safety, the application of HACCP can aid inspection by regulatory authorities and promote international trade by increasing confidence in food safety. For these reasons, the adoption of HACCP throughout the insect supply chain will be a determining factor in the success and development of the edible insect sector. According to FAO, "any HACCP system is capable of accommodating change, such as advances in equipment design, processing procedures or technological developments" (FAO/IAEA, 2001).

Although it has been stated that no significant health problems have arisen from the consumption of edible insects (Banjo, Lawal and Songonuga, 2006b), consumer confidence is arguably strongly correlated with the perceived safety of a given product. In this vein, the application of pesticides on insects destined for the food sector raises important issues, both for nutritional security and participation in the global market. It is well documented that species caught in fields, for example, are more likely to contain pesticides or heavy metals than those collected in dense forests. Chapulines (*Sphenarium purpurascens*) – red grasshoppers typically harvested in regions like Oaxaca, Mexico – have been found to contain high concentrations of lead from nearby mines (Handley, 2007). Many countries in Africa do not have policies governing the use of chemicals in fields in areas where villagers collect edible insects. Most of the time, collection takes place with little knowledge of the consequences that eating chemically treated insects might have (Ayieko *et al.*, 2012) (for more information on this topic, see Chapter 12). However, food-safety issues are important not only for insects collected in the wild but also for farmed insects.

10.1 PRESERVATION AND STORAGE

Insects are often consumed quickly after harvesting. Some insects are commercialized and transported within countries or beyond national borders for sale in distant markets; this is not uncommon between the Lao People's Democratic Republic and Thailand, for example. Live insects, after washing, are typically transported in ice coolers shortly after collection. Refrigeration is also recommended for fried and boiled insects.

Insects can be preserved and traded after (sun-)drying – a typical method used in processing the mopane caterpillar, for example (Allotey and Mpuchane, 2003) (Box 10.1). The dry environments typically found in places where sun-drying is common practice limits the growth of most micro-organisms. In humid areas, however, even sun-dried caterpillars are susceptible to moisture, which can stimulate the growth of microbes. Insects can also be re-contaminated during the drying process through air or soil; for this reason, hygienic practices during processing are of great importance and an additional heating/cooling step is recommended before consumption (Amadi *et al.*, 2005; Giaccone, 2005).

In many parts of the world, "ready-to-eat" insects are often sold in local markets after frying or roasting. In such cases, hygienic handling is equally important to prevent the potential risk of re-contamination and cross-contamination. At a household level, fresh insects should be prepared hygienically and sufficient heat treatment applied to ensure a microbiologically safe food product. Other simple preservation methods such as acidifying the insects with vinegar have been successful. Another example is the use of insects for protein enrichment in fermented food products. This is a viable processing option with mutual benefits, since the decreased pH in lactic acid-fermented products prevents the growth of potentially harmful micro-organisms (Klunder *et al.*, 2012).

There has been some success in processing and commercializing insects in the Netherlands. Three insect species (yellow mealworm larvae, lesser mealworm larvae and migratory locusts) can be found in specialized shops in the country that are produced and processed specifically for human consumption. One-day fasting is applied to ensure that the insect has an empty gut (degutting), and the insect is then freeze-dried whole. This produces a safe product with a relatively long shelf life (one year), if stored appropriately in a cool, dry place. Additional advantages of freeze-drying are the maintenance of the insect's nutritional value and the capacity of the product to re-absorb water. Nevertheless, obstacles remain: freeze-drying is expensive and often results in undesirable oxidation of

BOX 10.1

Processing the mopane caterpillar for human consumption

Care should be taken to avoid contamination throughout the various processing stages to ensure a safe product.

Degutting

- Ensure that holes have been dug for disposing of the gut contents.
- Holes must be covered immediately after degutting.

Drying

- Boil the bags (made of hessian or polypropylene) for at least 30 minutes and sun-dry for at least two hours before using them in the field.

Storage

- Ensure that bags are clean and disinfected before placing the caterpillars inside them.
- Ensure that bags are tied immediately with rope and the seams sewn up. They should then be covered with polythene material and placed on a raised platform to prevent cross-infestation from the surrounding environment and moisture from the ground from penetrating the bags.

Source: taken from Allotey and Mpuchane, 2003.

the long-chained unsaturated fatty acids, decreasing the nutritional value of the product and resulting in "off" odours and tastes.

A host of other contemporary preservation methods should be explored, such as the application of ultraviolet light and high-pressure technologies, as well as adequate packaging methods. Other important considerations need to be made in selecting the preservation method: the capacity to prolong shelf life (and in turn, contain costs), particularly if large amounts of insects need to be processed simultaneously; the extent to which the process preserves the nutritional value of the insects; and the cultural acceptability of the chosen preservation/processing method.

10.2 INSECT FEATURES, FOOD SAFETY AND ANTIMICROBIAL COMPOUNDS

Several issues related to food safety are distinct to insects because of their biological makeup: microbial safety; toxicity; impalatability; inorganic compounds; and the use of waste as insect feed.

Insect animal feed developed from manure and related organic waste streams raises bacteriological, mycological and toxicological concerns. Although some of these have been mentioned in the literature (Téguia, Mpoame and Okourou, 2002; Awoniyi, Adetuyi and Akinyosoye, 2004), they still have not been adequately researched (see section 5.2). The question is whether and to what extent insects sequester pathogenic organisms and toxic substances from manure and organic waste products.

10.2.1 Microbial safety

Insects may have associated micro-organisms that can influence their safety as food. Both insects collected in nature and insects raised on farms may be infected with pathogenic micro-organisms, including bacteria, virus, fungi, protozoa and others (Vega and Kaya, 2012). Such infections can be common. In general, insect pathogens are taxonomically separate from vertebrate pathogens and can be regarded as harmless to humans. Even within the genus *Bacillus* the insect pathogenic species *B. thuringiensis* and the vertebrate pathogen *B. anthracis* seem to have differing, non-overlapping life cycles (Jensen *et al.*, 1977). Also, insects have a high diversity of associated micro-organisms in their gut flora. Again, these organisms should, in general, not be seen as potential human pathogens. Finally, spores of various micro-organisms may be present on insect cuticles, including micro-organisms that grow saprotrophically on edible insect products and may even contribute to the degradation of the edible product. The above-mentioned micro-organism–insect association should, from the perspective of food consumption, be seen as microbial contamination and be treated as such.

In most tropical countries, insects are consumed whole, including their gut microflora. An exception is the mopane caterpillar, which is degutted (emptying the stomach by putting pressure on the body with two fingers), or fasted for one or two days before consumption. This process can affect the microbiological composition of an insect food product. Yet existing studies on the microbial safety of edible insects focus mostly on traditional practices of insect harvesting and consumption, making it difficult to decipher causal sources of infestations. Insect farming can allow greater control over hygienic practices and safe feed sources for insects, mitigating potential microbiological hazards.

The sanitary quality of the mopane caterpillar has been studied extensively given its frequent consumption in many African countries (Mpuchane, Taligoola and Gashe, 1996; Allotey and Mpuchane, 2003). One study carried out in Botswana demonstrated deterioration in sun-dried phane (the mopane caterpillar; *Imbrasia belina*) quality (i.e. disintegration of the inner flesh and change in colour due to mouldy growth and cavities in the chitinous exoskeleton). The most frequent fungal isolates found were species of *Aspergillus*, *Penicillium*, *Fusarium*, *Cladosporium* and *Phycomycetes*. Species or strains of *Aspergillus*, *Penicillium* and *Fusarium* are associated with mycotoxin production. Mpuchane, Taligoola and Gashe (1996) found levels of aflatoxins varying from 0–50 µg per kg of product; the maximum safe level set by FAO is 20 µg per kg. Frequent consumption of infected foods over longer periods is likely to pose health risks. Although the caterpillars in this particular study were degutted, boiled for 15–30 minutes and spread out on sheets or on the ground to sun-dry for 1–3 days, it was hypothesized that contamination was caused by one of the following sources: water of poor quality, insect vectors (such as flies and dipterans) and soil. To maintain the best sanitary quality, the study recommended drying the caterpillars quickly and evenly after harvesting and processing and storing them in a cool, dry place.

In West Africa, three rhinoceros beetle species of the genus *Oryctus* are commonly consumed: *O. monoceros* and *O. owariensis*, which breed in dead-standing coconut and oil palms, and *O. boas*, which occurs in rotting vegetation and manure heaps. Of the three beetle species, pathogenic bacteria were found in *O. monocerus*, including *Staphylococcus aureus*, *Pseudomonas aeruginosa* and *Bacillus cereus*, which may pose a risk to consumers (Banjo, Lawal and Adeyemi, 2006a). This contamination may have been due to improper processing, handling during retail and purchase, or exposure to air. As the consumption of the beetles contaminated with the pathogenic bacteria could pose risks to consumers, it was recommended that retailers of the partially dried and fried grubs ensure proper heating of the products to eliminate the pathogens.

The importance of hygienic handling and correct storage was highlighted by Klunder *et al.* (2012) in a laboratory experiment looking at the microbiological content of farmed yellow mealworm larvae (*Tenebrio molitor*) and house crickets (*Acheta domesticus*). Boiling the insects in water for a few minutes eliminated Enterobacteriacae, but spores were found to survive this process, with the potential that the spores could germinate and the bacteria grow given favourable conditions, such as temperatures around 30 °C and a moist environment, causing food spoilage. The spore-forming bacteria were found in the insect gut and on the skin and are likely to have been soil-borne. Alternative preservation techniques that do not involve the use of a refrigerator are drying and acidifying. Lactic fermentation of composite flour/water mixtures containing 10–20 percent powdered roasted mealworm larvae resulted in successful acidification and was demonstrated to be effective in safeguarding shelf life and safety by the control of enterobacteria and bacterial spores.

In another experiment, chemical–physical and microbiological analyses of the following five insect species with rearing potential were carried out: superworm (*Zophobas morio*), yellow mealworm (*Tenebrio molitor*), wax moth (*Galleria melonella*), butterworm (*Chilecomadia moorei*) and house cricket (*Acheta domesticus*). Neither *Salmonella* nor *Listeria monocytogenes* were identified in the analysed samples and it was concluded that it is unlikely that these insects attract microbial flora that pose risks to humans. However, it is still recommended that insects undergo a transformation to render inactive or reduce their microbial content. This could involve cooking (e.g. boiling or roasting) or pasteurization (Giaccone, 2005).

In contrast to being a potential microbial hazard, some edible insects are known to contain antibacterial peptides. A novel peptide (Hf-1) from the larvae of the common housefly (*Musca domestica*), for example, has been found to inhibit strains of food pathogens such as *Escherichia coli*, *Pseudomonas aeruginosa*, *Salmonella typhimurium*, *Shigella dysenteriae*, *Staphylococcus aureus* and *Bacillus subtilis*. The presence of Hf-1, also found in orange juice, suggests that the insect has potential as a food preservative (Hou *et al.*, 2007).

10.2.2 Toxicity

Some insect species considered toxic are eaten after precautionary measures are taken (Box 10.2). In Cameroon and Nigeria, *Zonocerus variegatus* has to be prepared in a specific way (Barreteau, 1999) by heating the insects in tepid water and then changing the water before cooking (Morris, 2004). Similarly, the tessaratomid *Encosternum* (=*Natalicola*) *delegorguei* in Zimbabwe and South Africa excretes a pungent fluid (Faure, 1944; Bodenheimer, 1951) that can cause severe pain and even temporary blindness if it comes into contact with the eyes (Scholtz, 1984). Therefore, the insect is consumed after removal of the fluid by squeezing the thorax and placing the bug in tepid water.

BOX 10.2

The stink bug *Nezara robusta* in southern Africa

Pentatomid bugs are widely consumed in southern Africa. Among these is the stink bug *Nezara robusta*, commonly found in *Brachystegia* woodland and blue gum (*Eucalyptus globulus*) plantations. The shield bug, another type of pentomid, is found at around 1 200 m above sea level in shady conditions and is associated with trees such as *Uapaca kirkiana*, *U. nitida*, *Brachystegia spiciformis* and *B. floribunda*. As its name suggests, the bug releases a powerful smell.

Women typically collect the bugs in the early morning because in cold conditions the cold-blooded insects are immobile and easy to collect. Collectors hit tree trunks with a log or long bamboo pole with a pouch fastened to one end to shake the bugs from the trees. The captured bugs are transported in bamboo baskets. They are prepared for consumption in tepid water, although care must be taken because the bug's bitter juice stains fingers brown and can be painful if it comes in contact with the eyes. After soaking in water, the bug ejects its bitter juices and loses its powerful smell. The bug is never placed in boiling water, as this would kill it immediately and cause the poison to be retained. After washing, the bug loses its green coloration and becomes a pale golden yellow. It is then cooked with a little water and salt. It is commonly eaten as a snack or side dish and the surplus sold in local markets. The water used to wash the bugs is said to be a useful pesticide against termites.

Source: Bodenheimer, 1951; Morris, 2004.

In the Carnia region of northeast Italy it is customary for children to eat the sweet ingluvies of a brightly coloured moth from the genus *Zygaena* (Zagrobelny *et al.*, 2009) (the ingluvies is the crop-widened portion of the oesophagus in many mollusks, insects and birds, which serves to accumulate, store and sometimes also begin the chemical processing of food). The moths contain cyanogenic glucosides, which release toxic hydrogen cyanide on degradation. They contain very low amounts of the toxic substance but high quantities of various sugars. Children collect the moths in early summer when they are plentiful and dissect them themselves, eating only the ingluvies.

There are limited reports, however, of adverse reactions caused by insect consumption. Cases of ataxia syndrome, characterized by tremors, ataxia and varying levels of impaired consciousness, were reported after consumption of the seasonal silkworm *Anaphe venata* in southwest Nigeria (Adamolekun, 1993; Adamolekun, McCandless and Butterworth, 1997). Further studies indicated that the reaction was most probably related to a structural undernourishment in the consumers who, being marginally thiamine deficient as a result of a largely carbohydrate-based diet containing thiamine-binding cyanogenetic glycosides, experience seasonal exacerbation of their thiamine deficiency from thiaminases in seasonal foods. In turn, they experience an adverse reaction to *A. venata*, which contains such thiaminases.

Edible insects can sometimes contain certain features that may be hazardous. For example, the consumption of caterpillars with hairs containing toxic substances can be very dangerous. These hairs have to be burned off (Muyay, 1981).

10.2.3 Unpalatability

In the Democratic Republic of the Congo, Bouvier (1945) observed that consuming grasshoppers and locusts without removing the legs caused intestinal constipation because the large spines on the tibia (shinbone) would catch in the gut. The only remedy in humans after consumption is often surgery to remove the legs from the gut. Similarly, in eastern Java, Indonesia, patients found to have eaten large quantities of roasted scarab beetles (*Lepidiota* spp.), whose indigestible chitinous remains can accumulate in several places inside the gut and cause total constipation, had to undergo surgery (Kuyten, 1960). Autopsies of dead monkeys following locust invasions revealed that the consumption of locusts proved to be fatal for the same reason. The product label of the migratory locust, Bugs Locusta, currently sold on the Dutch market, clearly states that the insect's legs and wings should be removed prior to consumption.

10.2.4 Inorganic contamination

Harmful metals from the environment have been found in the cells of several insect body parts – such as the fat, integument (exoskeleton), reproductive organs and digestive tracts – where they bio-accumulate. A study on the yellow mealworm larvae (*Tenebrio molitor*), for example, showed that the insects accumulate cadmium and lead in their bodies when they feed on organic matter in soils that contain these metals (Vijver *et al.*, 2003). However, Lindqvist and Block (1995) showed that after each moult, larvae lose some cadmium, and even larger amounts of the metal are lost after metamorphosis. Further research into the consequences this might have for human consumption is necessary.

Another issue of concern is the uptake of pesticides by edible insects such as locusts and grasshoppers, which can cause problems when they are consumed in large quantities. These risks are of major concern in the traditional practices of harvesting and consuming insects in the wild, where the control of chemical applications is difficult (Box 10.3). This is yet another potential benefit of insect rearing, where chemical hazards can be controlled to a larger extent.

BOX 10.3

Bogong moths in Australia

Each spring, the cutworms of the bogong moth (*Agrotis infusa*) leave the lowland breeding grounds of eastern Australia to escape the harsh summer environment. They migrate up to 1 000 km to the Snowy Mountains and the Victorian Alps in southeastern Australia, spending the summer in large congregations in caves and crevices between rocks at about 1 200 m above sea level. In autumn, they make the return journey home. During this migration, they are often considered a nuisance because they are attracted to lights on houses and other buildings at night.

Indigenous Australians are documented to have once harvested the moths from the Alps using the heat and smoke of torches, a custom no longer practised. They feasted on this tasty, high-protein and high-fat resource. Yet a study by Green *et al.* (2001) showed that the insects transport sub-lethal quantities of arsenic taken up in the lowland breeding grounds from plants in areas where herbicides have been used. Arsenic was concentrated in damaging levels on the aestivation (similar to hibernation) sites on the Alps, as a result of the millions of moths congregating in these areas. The arsenic was also detected in soil inside the caves and on grass from outwash areas.

Source: Green *et al.*, 2001.

10.3 ALLERGIES

10.3.1 Allergic reactions to edible insects

Like most protein-containing foods, arthropods can induce allergic reactions in sensitive humans (Immunoglobulin E (IgE) mediated). These allergens may cause eczema, dermatitis, rhinitis, conjunctivitis, congestion, angioedema and bronchial asthma. While some people have a history of atopy (allergic hypersensitivity), it is also possible to develop allergic sensitivity through long-term exposure. The majority of cases are inhalant or contactant in nature (Phillips and Burkholder, 1995; Barletta and Pini, 2003). Allergic reactions to bee and wasp venom (injectant allergens) are well known.

Individuals in constant contact with insects, such as entomologists, laboratory workers (working mostly with beetles, cockroaches, locusts, blowflies, crickets, moths or flies) and agricultural and industrial workers (working mostly with bean weevils, grain weevils, mushroom flies, sewer flies, houseflies, silkworms or fish bait, such as the larvae of flies and moths) are most vulnerable to such allergies. Ways of developing allergic reactions include inhalation of dust containing cockroach fecal matter, and skin contact with caterpillar hairs. Studies suggest that people frequently in contact with larvae of *T. molitor*, for example, run the risk of developing certain allergic reactions (Senti, Lundberg and Wüthrich, 2000; Siracusa *et al.*, 2003). The same was found for the closely related species *Alphitobius diaperinus*. The symptoms of the allergic reactions include inflammation of the eyes and nose (*T. molitor*) and itching, mild swelling, inflammation of the nose, asthma and skin rash (*A. diaperinus*) (Schroeckenstein *et al.*, 1988; Schroeckenstein, Meier-Davis and Bush, 1990). Cross reactivity can also occur between the two species, meaning that the antibodies for a specific allergen in one insect species is capable of identifying allergens in another and may thus induce an allergic reaction to that insect as well. Cross-reactivity is not absolute, however; some people develop allergic reactions to specific insects with little cross-reactivity to other insects due to long-term exposure to high amounts of allergens from that specific insect. In household settings where several insects and other arthropods can co-occur, it is difficult to assess whether an allergic person has multiple sensitivities caused by all arthropods or a general allergic sensitivity to invertebrates (cross-reactivity) (Barletta and Pini, 2003). Tropomyosins (actin-binding proteins that regulate muscle contractions) from cockroaches, mites and shrimps have been reported to be allergenic. Some patients allergic to dust mites that were increasingly exposed to mite antigen became sensitive to seafood tropomyosins, for example (Reese, Ayuso and Lehrer, 1999). These findings suggest that people with seafood allergy, for example, could experience allergic reactions to the consumption of edible insects.

There is a certain amount of evidence of allergies induced through the ingestion of insects. Because honeybee larvae contain pollen, for instance, people allergic to pollen are advised not to eat them (Chen *et al.*, 1998). Asthmatic symptoms were recorded on ingestion of *Orthoptera* (Auerswald and Lopata, 2005). In one study conducted in the Lao People's Democratic Republic, one respondent with a history of consuming insects is said to have developed an allergy to giant water bugs, while another is said to have developed an allergy to all edible insects, as well as shrimps (J. Van Itterbeeck, personal communication, 2012). This raises the question of the potential of developing sensitivity caused by ingesting edible insects and by handling while cooking and eating. It is doubtful whether processing measures such as boiling will destroy allergenic components (Phillips and Burkholder, 1995). For the great majority of people, however, eating and/or exposure to insects do not pose significant risk of causing allergenic reactions, especially if the individuals have no history of arthropod or insect allergen sensitivity acquired through long-term exposure to an allergen in sufficient quantities.

BOX 10.4
The allergy–hygiene hypothesis[13]

Allergies are an increasing problem in Western populations, contrary to developing countries where their prevalence is far lower. The hygiene hypothesis states that the high prevalence of allergies in Western populations is induced by a lack of exposure to pathogens, including intestinal parasites, and to increased vaccination practices during childhood.

Most parasites contain chitin. It is hypothesized that the variation in exposure to chitin and to intestinal parasites may be a key to explaining the asymmetric prevalence of allergies in populations. The presence of chitinases in human gastric juice has been associated with responses to parasitic infections and linked to allergic conditions. A review of the immunological response to chitin and its possible role in inducing asthma and allergies revealed that the responses appeared to depend on the particle size of the chitin substance; in other words, medium-sized chitin particles induce allergic inflammation, while small-sized chitin particles may have the reverse effect of reducing the inflammatory response (Brinchmann *et al.*, 2011). The consequences for the pathogenesis of asthma and allergies following increased consumption of chitin through the promotion of insects as food are unpredictable. However, if allergies are catalysed by a lack of exposure to chitinous substances in childhood, as suggested, increasing the consumption of insects in early childhood could, by extension, support better protection against allergies later in life.

[13] N. Roos contributed this box.
Source: FAO/WUR, 2012.

10.3.2 Immunological effects of chitin, a major component in insect cuticle

Chitin, the second most abundant polysaccharide in nature, contains nitrogen and is commonly found in lower organisms such as fungi, crustaceans (e.g. crabs, lobsters and shrimps) and insects, but not mammals. Although the anti-viral and anti-tumour activities of chitin/derivatives have been known for some time, the immunological effects of chitin have only recently been recognized (Lee, Simpson and Wilson, 2008). Recent studies have demonstrated that chitin has complex and size-dependent effects on innate and adaptive immune responses (see Lee, Simpson and Wilson, 2008). In several studies, it was suggested that chitin is an allergen (Muzzarelli, 2010). However, chitin and its derivative, chitosan (produced commercially by the de-acetylation of chitin), rather than acting as allergens have been found to have properties that could improve the immune response of specific groups of people (Goodman, 1989; Muzzarelli, 2010; H. Wichers, personal communication, 2012) (Box 10.4). By inducing non-specific host resistance against infections by pathogenic bacteria and viruses, there are indications that chitin reduces allergic responses in individuals. Moreover, chitin has shown potential for boosting immune system functioning, making it a promising alternative to antibiotics currently used in livestock (H. Wichers, personal communication, 2012). The use of chitin for medical and industrial purposes needs to be explored further.

11. Edible insects as an engine for improving livelihoods

For most people living in rural areas, especially the poor, forests and trees are important sources of food and cash income. Some 350 million of the world's poorest people – including 60 million indigenous peoples – depend on forests for their daily subsistence and long-term survival (FAO, 2012a). Insects are a major source of animal protein in many communities and are critical for diet diversification but, in most countries, eating insects is not a matter of survival but a question of personal choice. In fact, the vast majority of insect consumption is by choice, not necessity, and insects are a part of local culture. Nevertheless, insects do provide valuable buffers against seasonal shortages of food (Dufour, 1987). As well as acting as important food items, insects provide additional cash for basic expenditure, including on food, farming inputs and education (Agea *et al.*, 2008; Hope *et al.*, 2009).

Trade in edible insects is a major source of income in some places and what is collected is not always fully consumed directly by the collectors themselves. Insects offer important livelihood opportunities for many people in developing countries, including some of the poorest segments of society and particularily women and children. In a post-Rio+20 context, "greening" the economy with forestry – including edible insects – can help redress the social, economic and regional asymmetries and inequalities that still prevail in many parts of the world (FAO, 2012d). This chapter looks specifically at the potential of insects to improve local diets, contribute to strengthening access and tenure rights to local resources, and provide opportunities for improving the livelihoods of women. The economic contribution (sale and cash income) of insects is presented in Chapter 12.

11.1 INSECTS AS PART OF THE MINILIVESTOCK SECTOR

The livestock sector is important in providing income and livelihoods for people around the world and accounts for 40 percent of agriculture's total gross domestic product (Steinfeld *et al.*, 2006). However, despite the growing demand for animal products, livestock ranging beyond classical species such as pigs, goats and chickens are often not seen as significant. However, minilivestock (defined in Chapter 4), such as insects, can be important for economic diversification.

Insect cultivation can be carried out in urban, peri-urban and rural areas and is an efficient use of space (Oonincx and de Boer, 2012). Although some attempts have been made to domesticate certain insects, most species, such as tarantulas, can only be collected in the wild (C. Munke, personal communication, 2012). Minilivestock enterprises are advantageous because they (FAO, 2011b):

- require minimal space;
- do not compete directly with food for human consumption;
- have a demand which outstrips their supply;
- have high reproductive rates;
- create cash inflow in a short period;
- have high to very high financial returns in many cases;
- are nutritional and a part of human nutrition;
- convert feed to protein efficiently;
- are relatively easy to manage;
- are easily transportable;
- are often easy to raise and do not require in-depth training.

Insects, along with other minilivestock, support diversified markets because they can be sold to consumers across the rural–urban spectrum. In many cases, rural people will sell their minilivestock within their villages; however, due to their transportability, insects can easily be moved to urban markets by, for example, bus, truck or bicycle. Rearing insects can also be carried out as a complement to other livelihood strategies. Additionally, insect rearing can be done by both landowners and the landless (FAO, 2011b) because not much space is required.

For domestic insect production, insects are reared in a controlled setting (farming), and the assurance of consistent production can evenly distribute related annual earnings, an advantage over the seasonality of some insects. For many households, a consistent income flow promotes savings and the ability to pay periodic expenditures such as school fees (FAO, 2011b). However, the wild collection of insects should not be considered insignificant because it can also contribute to livelihood diversification. In some regions of Africa, for example, the consumption of insects has been estimated to fluctuate between 2 percent and 30 percent of total meat consumption in a year, depending on the availability of insects (FAO/WUR, 2012).

In many cases, rearing minilivestock is not resource intensive; thus, it can be practised by women, men, the elderly and even children. Insect rearing and collecting, in both captivity and the wild, generally requires only a small initial investment followed by small, incremental investments, which can reduce economic risk. Infrastructural investments can include nets, plastic sheets and containers (Hanboonsong, 2010). Therefore, insect rearing and harvesting can contribute positively to equal participation and involvement in economic growth, especially for marginalized groups such as the landless.

11.2 IMPROVING LOCAL DIETS

Most insects collected by rural people are destined for personal consumption; however, this depends significantly on location and species. Any excess is likely to be sold in local and regional markets. In Papua, Indonesia, for example, indigenous communities consume between 60 and 100 edible insect species to diversify their dietary needs. Surplus is sold in markets, such as the black palm weevil (*Rhynchophorus bilineatus*), which is collected from the sago palm (*Metroxylon sagu*). Palm weevils are the most commonly consumed insect species in the region (Box 11.1). The price of one bag containing 100–120 larvae fetches US$2.11 in local markets, comparable in value with 20 chicken eggs and 3 kg of rice (Ramandey and Mastrigt, 2010). However, some insects are collected more for commercial sale than for household consumption. Moreover, there is a lot more commercial, albeit often informal, trading of insects than is generally acknowledged or documented.

Although edible insects are often consumed for their taste, they can also provide food in times of low supply. Nonetheless, it should be emphasized that the consumption of insects as a source of emergency food is not the most common reason for their consumption. A survey conducted in Kinshasa in 2003 reported that 70 percent of the city's population eats mopane caterpillars because of their nutritional value and taste. However, during periods when other protein sources may be more difficult to find, insects are an important part of the diet. In the Central African Republic, 95 percent of forest people are dependent on eating insects for their protein intake (FAO, 2004). Insects are sometimes the only source of essential proteins (amino acids), fats, vitamins and minerals for forest people.

In many parts of the world, consumption is seasonal for two reasons: local people feed on seasonally available plants; and the surplus is not stored due to a lack of processing and conservation methods. In West and Central Africa, as well as parts of the Amazon, bushmeat and fish are scarce during the rainy season, and it is during this time that edible insect consumption increases considerably (see Chapter 2). Not surprisingly, insects tend to be more abundant during the rainy season, when they constitute valuable parts of local diets and provide essential nutrients.

BOX 11.1

The red palm weevil (*Rynchophorous ferrugineus*) as an important source of nutrition and livelihood in New Guinea

Rural people on the island of New Guinea (Papua New Guinea and Indonesia) have developed an intimate relationship with insects. Many species form an integral part of subsistence diets: beetle larvae and adults (Cerambycidae, Scarabaeidae and Curculionidae); cicadas (*Homoptera*); stick insects (*Phasmida*); termites (*Isoptera*); mayflies (*Ephemeroptera*); wasp larvae (*Hymenoptera*); caterpillars and moths (*Lepidoptera*); dragonfly larvae (*Odonata*); grasshoppers and locusts (*Orthoptera*); and many spider species (*Arachnida*). However, the larvae of the red palm weevil (*Rhynchophorus ferrugineus papuanus*), which grows in the trunk of the sago palm, is the most widely consumed insect on the island. Special festivities are organized at which many palm trees are cut to collect weevil larvae. Raised as a byproduct of sago starch (prepared from carbohydrate material stored in the trunks of several palms, including *Metroxylum rumphii*), the larvae are commonly found in markets around the island. One study documented how at the market in Lae, Papua New Guinea, village women sold about 40 larvae (250 g) for approximately US$1.00. Buyers will pay about US$0.50 in the market for 12–15 larvae when they are grilled.

Locals grill, boil and roast the larvae and sometimes fortify sago pancakes with the insects, increasing the pancakes' nutritional value. In parts of the island where sago is a staple, the consumption of the red palm weevil provides locals with much-needed protein, as sago starch is low in protein. According to WHO, every 100 g of weevil larvae contains: 182 kilocalories, 6.1 percent protein, 13.1 percent fat, 9 percent carbohydrates, 4.3 mg iron, 461 mg calcium, and other vitamins and minerals.

Source: Mercer, 1997.

11.3 ACCESS, TENURE AND RIGHTS TO NATURAL CAPITAL

Access to natural resources is an important factor in the improvement of livelihoods. Insect harvesting can help to strengthen tenure rights and increase responsibility for the conservation of natural resources.

Insects provide an easily accessible source of income in many rural areas, particularly for women and children who are typically involved in their harvest. Insects can be directly and easily collected from nature – with minimal expenditure (i.e. basic harvesting equipment) – when access to land (farmland or forests) to harvest insects is not limited. For vulnerable segments of society, like indigenous people, women and the elderly, access to land is a traditional impediment to livelihood development and could thus present a barrier to edible insect harvesting. Insect collection can generally take place in public forests. This makes the practice far more accessible than many traditional agricultural activities that require either direct access to land or land tenure. Insect harvesting from nature may be less destructive to forest resources than the gathering of NWFPs such as medical plants or rattan, which require killing the host plant. For this reason, insect harvesting should be recognized as a vital component in ensuring food security, as long it is performed in a sustainable manner.

The reduced availability of wild edible insects is set to make collecting more difficult and in turn lead to lower consumption and trade in insects. For this reason, conservation and management measures need to be put in place to protect insects and their environments (Yhoung-Aree, 2010) (see section 4.3). Local authorities should recognize the contribution that insect harvesting makes to the livelihoods of local people. Once local people see the benefits that can arise from the participatory management of natural resources, they may be more convinced to protect the (forest)lands in which the insects are gathered and more eager to participate.

While overexploitation and overharvesting are concerns, there are few documented cases in which collection has depleted arthropod populations (Box 11.2). In some cases of claimed overharvest, it has later been realized that declining insect populations were a part of natural population fluctuations and cycles. There is also a risk that, since farming can produce larger volumes than collecting, it will displace poorer collectors from their livelihoods.

BOX 11.2
Cambodian spiders

The sale of spiders is an important source of income for many poor farmers in Cambodia, who average a daily income of US$2. A species of tarantula, *Haplopelma albostriatum* (Thai zebra tarantula), locally named a-ping, is typically served fried and sold in street stalls in Skuon at the Kampong Thom market or in restaurants in the capital, Phnom Penh. Spiders are collected from forests or in cashew nut plantations during the day. Vendors buy live spiders from collectors, who find and dig them out of their dens, selling up to 100–200 spiders per day. There are about twelve vendors in Skuon alone. The fear is that these spiders will be collected to extinction. Vendors are reporting a sharp decline in numbers and blame farmers for clearing and burning forests (Yen, Hanboonsong and van Huis, 2013).

11.4 INCLUSION OF WOMEN

Rural communities across the developing world – especially vulnerable segments of society, such as women and indigenous peoples – depend greatly on natural resources, including insects, which act as a buffer against poverty (Box 11.3). In South Africa, for example, research on the use of a range of bioresources among 110 households in Limpopo Province found that the use of natural resources including wild herbs and fruit as well as edible insects was extensive among poor households (Twine *et al.*, 2003). However, access to natural resources is sometimes restricted for historical and cultural reasons. For example, although many countries have extended legal rights to women over land inheritance, customary practices as well as the inability of women to assert those rights makes ownership of land problematic. Ensuring equitable access to local natural resources and, by extension, to wild foods, including edible insects, remains a key factor in ensuring food security.

Women are the backbone of rural economies, especially in the developing world. Yet they still encounter difficulties in accessing essential resources such as land, credit, inputs (including improved seeds and fertilizers), technology, agricultural training and information. Studies show that empowering and investing in rural women can significantly increase productivity, improve rural livelihoods and reduce hunger and malnutrition. It is estimated that if women had the same access to productive resources as men, their farm yields could increase by 20 percent or even 30 percent. Moreover, closing the gender gap in agriculture could lift 100–150 million people out of hunger (FAO, 2011c).

Throughout the world, many women are engaged in small and medium-scale forest-based enterprises and depend on forest products for generating income. They are actively involved in collecting, processing and marketing a number of NWFPs, including edible insects. One study found that over 94 percent of the 1 100 NWFP traders surveyed in rural and urban markets in Cameroon were women. The same study showed that in the Democratic Republic of the Congo, more women than men participated in the bushmeat trade, representing 80 percent of bushmeat traders in Kinshasa markets (Tieguhong *et al.*, 2009). Yet for most of the time these activities are informal in nature, for a number of reasons: women tend to have greater household responsibilities, which limit their ability to participate fully in formal economies; women often have unrecognized or low levels of skills and education; and any income derived from the sale of NWFPs tends to be used for household needs rather than for expanding their businesses (FAO, 2007). Small and medium-sized forest-based enterprises provide an opportunity for the edible insect sector to reduce poverty, improve equity and protect forests and other natural resources.

Women and children play active roles in the edible insect sector, mainly because the entry requirements to engage in insect collection, processing and sales are relatively low. In southern Zimbabwe, the collection, processing (removing gut content, roasting and drying), packing, blending and trading of mopane caterpillars have traditionally been carried out by women (Hobane, 1994; Kozanayi and Frost, 2002) (Box 12.1). Women are the main sellers of mopane caterpillars in towns and small business centres, mostly in small volumes (Kozanayi and Frost, 2002), but men tend to dominate the more lucrative long-distance and large-volume trading chains. The main problem cited by the women is that the large volumes of mopane caterpillars are too cumbersome to transport to make cross-border trade worthwile. For these reasons, women generally sell their catch in small volumes at open markets, sales points along roads, bus termini and municipal markets. Most women collectors and processors come from local communities and are traditionally highly immobile. They also have many domestic obligations to fulfil, such as working in the field, harvesting food, cooking, looking after the children, and collecting fuelwood and water.

In Mexico, studies have shown that gender plays a significant role in the search, collection, preparation, marketing and sale of edible insects among ethnic groups (Ramos Elorduy, Carbajal Valdés and Pino, 2012). Women and children tend to be the principal foragers if the species in question is relatively easy to access. Poisonous insects, and insects that inhabit dangerous environments, are generally harvested by men. Additionally, while insects harvested by women generally contribute to household food needs, men's harvests typically end up on the wholesale market, especially when large volumes have been collected. Women assist men in this process by selling insects on the retail market. Insects sold by women include grasshoppers, stink bugs (jumiles), the giant mesquite bug (*Thasus giagas*) (xamues), small beetles, cicadas, the immature larvae of butterflies and moths, ants (*Atta* spp., including their queens, chicatana) and stingless bees. Not surprisingly, honey derived from stingless bees is the best-selling insect product at local markets (Ramos Elorduy, Carbajal Valdés and Pino, 2012).

BOX 11.3

Edible insect consumption and indigenous peoples

Indigenous peoples live in symbiosis with their natural environment and are highly dependent on natural resources for their livelihoods. For this reason, they have time-hounoured understandings of how and where to find insects and different methods of preparation. This knowledge is particularly important in times of food shortages (Ramos Elorduy, 1984).

In Australia, "bush foods", including edible insects, are highly valued by Aborigines. They are still harvested and seen as an integral part their culture. Some of the better-known insects that they traditionally consume are edible beetle larvae and caterpillars (witchetty grubs), honey ants, scale insects, lerps and the Bogong moth, *Agrotis infusa* (Yen, 2005).

Continues

Box 11.3 continued

In addition to caterpillars, honey ants have traditionally been an important part of local cultures, and are generally gathered by women and children (Yen, 2010). Honey ants were important seasonal sources of carbohydrates for indigenous Australians and also serve as a living food store for other ants in a colony. They hang from the ceiling of underground chambers and are gorged with food from other workers. The food is stored in the abdomen, which can become distended to many times its normal size. The ant remains hanging, sometimes for months, until the ant colony needs the stored food. After stimulation, the ant regurgitates the sweet honey.

Given their different activities, men, women and children naturally possess different types of knowledge about insects. One study in Niger found that women were able to name as many as 30 grasshopper species using their vernacular names, which was approximately ten more than men (Groot, 1995), because women play a larger role in collecting and preparing the insects. Similarly, Aboriginal men and women contribute differently to subsistence diets in Australia (women providing plants, honey, eggs, small vertebrates and invertebrates and men primarily hunting larger vertebrates) and consequently their knowledge varies vastly (Yen, 2010). Given the importance of traditional ecological knowledge in improving general understanding of insect ecology and biology, policies on the sustainable management and development of the edible insect sector must take into account the different roles that men and women play in edible insect practices, and involve them accordingly.

For a number of reasons, protein and other nutritional deficiencies are typically more widespread in disadvantaged segments of society. Women and other vulnerable people are at a disadvantage in accessing productive resources. Women have different biological needs than men, necessitating a more targeted dietary regimen. For example, women typically need 2.5 times more dietary iron than men, as well as more protein when pregnant or lactating (FAO, 2012e). Access to non-insect animal protein also differs between men and women in some societies, with men generally having greater access. For the Tukanoan Indians in the northwest Amazon, for example, insects provide up to 12 percent of the crude protein derived from animal foods in mens' diets in one season, compared with 26 percent in the women's diets (Dufour, 1987). The same study provided examples showing that insects are the only protein source accessible to women during certain periods of the year; at the same time they constitute important sources of fat. Because of their nutritional composition and the relative accessibility of the resource, edible insects offer an important opportunity to counter nutritional insecurity and improve livelihoods among vulnerable people.

12. Economics: cash income, enterprise development, markets and trade

This chapter illustrates key economic aspects of insect gathering and farming, such as their potential to produce income at the household level or at larger, industrial scales. It presents key characteristics for developing insect-based enterprises, including how to bring insects to market, and provides data on trade in edible insects.

12.1 CASH INCOME

Gathering and/or farming insects can offer unique employment and income-earning opportunities in developing countries, particularly, but not exclusively, for the poor in urban and rural areas. In many cases, insect collection and cultivation can serve as a livelihood diversification strategy that provides multiple income-generating opportunities for households. For example, silkworms, ants and bees can be considered as multipurpose production systems: silkworms can be used for food and fibre, and weaver ants (*Oecophylla* spp.) combat pests and can be used as food (Offenberg and Wiwatwitaya, 2009b). In the case of bees, both the honey and the larvae can be harvested as food. For example, the Hazda foragers of Tanzania do not remove the bee larvae from the combs when eating the honey (Murray *et al.*, 2001).

In developed countries, insect rearing occurs mainly on family-operated farms. Presently, there are only a few large-scale industrial plants that rear insects. This section looks mainly at the income-earning potential that insects offer poor people in developing countries. The income-earning potential of rearing and processing insects in developed countries is described in following sections.

The large majority of insects used for food in developing countries are gathered from wild populations in nature, on farmlands or in forests. In addition to self-consumption by the gatherer and his family, the excess production of gathered insects can easily be sold for cash (or bartered) at local village markets or in street food stalls by the gatherer or their family members. As well as being sold directly to consumers at local markets, insects can be sold to middlemen and wholesalers at the farm gate. Their interaction and the number of middlemen involved, if any, will set the final price of the insect product for the end consumer.

Examples of insect prices are shown to give an idea of orders of magnitude and are taken from a wide variety of specific cases at the village level, in markets and online. These prices should not be used for extrapolation and should not be taken out of context. In Kenya, 1 kg of termites sells for €10 (V. Owino, personal communication, 2012). One can purchase 70 g of weaver ant pupae online for €7.50 in the United Kingdom of Great Britain and Northern Ireland. In the Netherlands, 50 g of the yellow mealworm and the lesser mealworm costs €4.85, and 35 migratory locusts cost around €9.99 online. In the Lao People's Democratic Republic, the price of grasshoppers is much lower, at approximately €8–10 per kg. In Oaxaca, Mexico, chapulines sell for around €12 per kg. At markets in Cambodia, one can (150–200 g) of fried crickets (*Acheta testacea* and *Gyllus bimaculatus*) sells for €0.40–0.70. Prices vary between rural and urban areas (C. Munke, personal communication, 2012).

In Thailand, middlemen buy insects from farmers to sell as food to wholesaler buyers, who then distribute the products to street vendors and/or retailers. Middlemen can also

BOX 12.1

Harvesting, processing and trade of mopane caterpillars

Women and children perform much of the work involved in harvesting and processing mopane caterpillars. The following tasks are necessary to produce a saleable product.

Collection. Women and children collect the caterpillars, preferably from short trees. When fully grown the caterpillars come down from the trees and are also collected on the ground.

Removing the gut contents. Fully grown larvae empty their guts before pupation. Most consumers prefer larvae at this stage. However, if not fully grown, the larvae need to be squeezed to remove the frass from the guts. The conventional way is squeezing between the thumb and forefinger. Some processors use a bottle as a roller to expel the frass. Buyers confirm whether the guts are clean by breaking the larvae in half. The spines on the larvae can puncture the hands of workers, causing discolouring, bleeding and sores. Some collectors tie bark fibres around their fingers or use gloves.

Roasting and drying. The processed larvae are roasted on smouldering charcoal to cook the caterpillars and remove the spines. Roasting also removes red colouration. Buyers check for signs of colouration to ensure that the larvae have been properly roasted. The caterpillars are then dried in the sun. Some people just add salt and dry them in the sun without roasting. Caterpillars prepared in this manner are only sold in local markets, as outsiders tend to prefer them without spines. Moreover, consumers in urban markets do not want salted mopane caterpillars because they have an undesired whitish appearance. The larvae may be boiled and then dried in the sun. As with salted and sun-dried caterpillars, boiled caterpillars still have spines, which reduce their market value.

Packing and blending. The caterpillars are packed in sacks or large tins to sell to traders or at market. Traders buy and repack the caterpillars in small packets before re-selling. Traders who buy mopane caterpillars in bulk from collectors usually mix low-quality larvae (e.g. those not properly squeezed out) with better-quality ones. This is called blending. A 35-litre bucket contains about 10 kg of mopane caterpillars.

Sales. Buyers rather than sellers determine the prices in most areas. The transaction often involves barter trading and the rate of exchange fluctuates considerably.

Transport and trading. Mopane caterpillars are often sourced far from the main markets, most of which are in urban areas. Traders often have to travel long distances to do business. Transport costs per unit of mopane caterpillars (travel cost plus freight cost) decrease as the weight of product being transported increases.

The traders. Women are confined largely to collecting and processing mopane caterpillars and to selling small volumes at open markets, sales points along roadsides, bus termini and municipal markets. Men generally dominate the more lucrative long-distance and large-volume trading chains of mopane caterpillars.

Markets. Mopane caterpillars are sold at a wide variety of outlets, both to consumers and to other traders. The major outlets are supermarkets and stores, bus termini, open municipal and roadside markets, and beer halls. Supermarkets are the main retail outlets for pre-packed and labelled mopane caterpillars supplied by wholesale food-packing companies

International trade. At the time of the survey, mopane caterpillars from southern Zimbabwe were being traded to Botswana, the Democratic Republic of the Congo, South Africa and Zambia. Details of this trade are difficult to come by because of its informal and, in some cases, illegal nature (e.g. avoiding customs duty).

Seasonal fluctuations in availability and price. Local fluctuations in price throughout the year largely reflect variations in supply and demand. Some traders stockpile mopane caterpillars to sell during shortages, when prices are higher.

Source: Kozanayi and Frost, 2002.

provide live insects as breeding stock for sale to farmers who start rearing operations (A. Yen, personal communication, 2012).

Consumers can buy processed and unprocessed insects at village markets, retail supermarkets and stations and from street vendors. Insects can also be consumed in restaurants, depending on the extent to which entomophagy is recognized in a given region.

In southern Zimbabwe, mopane caterpillars are sold at rural and urban markets, and several market players are involved (Box 12.1). In Thailand, both fresh and cooked edible insects are sold at local markets, wholesale supermarkets and minimarts (Box 12.2). They are available either precooked (at street carts and food stalls) or uncooked in frozen packages in supermarkets (Hanboonsong, 2012). They are also marketed in ready-to-eat and microwaveable packages.

BOX 12.2
Wholesale markets in Thailand

Well-known edible insect wholesale markets are Rong Kluea at Sakeaw Province (the biggest edible insect market near the Cambodian border), Klong Toey market (Bangkok), Talad Thai (Bangkok), and Jatujak market (Bangkok; this market sells mainly mealworms as edible insects for pet feed) (Hanboonsong, 2012). With increasing demand for edible insects, quantities of these insects are also brought into Thailand from Cambodia and the Lao People's Democratic Republic to sell at the Rong Kluea market. These imported edible insects are mainly collected in the wild.

12.2 ENTERPRISE DEVELOPMENT

People gathering insects in the wild for sale at nearby markets generally act on an individual basis in the same way as insect farmers. Box 12.3 summarizes the key feasibility issues to be taken into account when starting an insect-based enterprise.

BOX 12.3
Feasibility study before starting a street-food business

Market feasibility:
- types of street and snack foods sold (are insects already on sale?);
- selling prices of street and snack foods;
- types of customers (families, children, office workers, etc.);
- buying frequency of customers;
- quantities sold;
- competition;
- quality and safety required by customers.

Technical feasibility:
- processing and preparation methods required to provide the desired quantities;
- hygiene and safety requirements for processing;
- legal requirements for hygiene and safety;
- farm produce required to supply ingredients;
- equipment needed;
- labour needed;
- skills required.

Continues

Box 12.3 continued

Financial feasibility:

- start-up costs;
- operating costs;
- cash flow;
- profit potential;
- loans.

Source: FAO, 1997a.

12.2.1 Formation of cooperatives and associations

Insect product-based enterprises are still very much an emerging value chain. This means that there will be challenges, such as in legislation and the regulation of the edible insect sector, which cannot be solved by individuals. Therefore, stakeholders need to work together to further their common agenda, strengthen recognition for their activities and increase their bargaining power.

Associations and organizations are an essential link between decision-makers, NGOs and farmers. By offering a common voice to farmers and gatherers – in this case those who are involved in the edible insects sector – they can aid the planning, design and implementation of policies and programmes that directly or indirectly affect their livelihoods (FAO, 2007). In short, associations of insect producers (gatherers and/or farmers) may prove a powerful tool for the development of the sector. Box 12.4 provides an example of such an association.

BOX 12.4

The Dutch Insect Farmers Association

The production and sale of edible insects in the Netherlands started with the foundation of the Dutch Insect Farmers Association (VENIK) in 2008. Acknowledging the cultural barriers at play, the association opted for a long-term strategy that focuses on insects not only as food but also as feed and pharma. However, designing a future with insects as food requires action through lobbying, the development of business scenarios and roadmapping. VENIK is building a network at the national and international levels with market parties, knowledge institutions and NGOs. It has contacts with policymakers, politicians and the food safety authority. It also provides information on edible insects to professionals, consumers and the media.

VENIK maintains the hope that insects will one day be considered a nutritious, sustainable and credible source of protein. Special production lines are already in place to comply with HACCP standards. Three insect species are being produced for human consumption: the yellow mealworm (*Tenebrio molitor*), the lesser mealworm (*Alphitobius diaperinus*) and the migratory locust (*Locusta migratoria*). These insects are sold freeze-dried.

In the last few years, VENIK has been involved in developing legislation, quality standards and markets. The association is also building a knowledge base to promote acceptance and technical innovation at three levels: substantiation of insects as a protein replacement and techniques for automation; applying the knowledge gained; and performance of practical experiments.

Source: FAO/WUR, 2012.

Organizations can act as an alternative form of support when private and public services have failed. Moreover, they are free to operate in their own manner (FAO, 2007). Several producers at the Expert Consultation Meeting on Assessing the Potential of Insects as Food and Feed in Assuring Food Security at FAO in January 2012 issued a call for an international association of insect producers. Successful associations need to be market-oriented and effectively managed and must have a good organizational structure that clearly sets out everyone's rights and responsibilities, caters for the needs of all members, considers gender issues, and allows for freedom of speech. The benefits of creating and joining such organizations include (FAO, 2011b):

- reductions in the costs of buying farm inputs, production, processing and marketing;
- sharing and pooling resources and skills and the acquisition of new skills as a consequence of cooperation;
- lower transaction and transport costs;
- improved access to credit;
- improved capacity to access urban areas;
- more opportunities for market linkages;
- greater opportunities for training in hygiene, food preparation, business skills development, etc.;
- a unified voice for obtaining licences and permits for marketing from authorities;
- improved social cohesion among members.

12.2.2 Example of strategies for enterprise development: learning from other industries

It is estimated that sericulture can generate 11 workdays of employment per kg of raw silk production (in on-farm and off-farm activities). No other industry generates this level of employment, especially in rural areas; hence, sericulture is used as a tool for rural reconstruction. Sericulture also provides vibrancy to village economies – about 57 percent of the gross value of silk fabrics flows back to the rural cocoon growers (Umesh *et al.*, 2009). The Government of India's XI Plan (2007–2012) stated that extra research and development should be conducted to heighten production capacity, infrastructure development, human resources and other facilities. Production, employment and export rates all grew in 2010–2011 (Government of India, 2011).

The 2011 annual report of India's Ministry of Textiles (Government of India, 2011) provides details of marketing strategies for sales of wild "Vanya" silk products on international markets. The Vanya silk market promotion cell focuses on the promotion, product development and diversification of this type of silk, and has published a directory of manufacturers, traders, retailers and exporters. The promotion cell also organizes and participates in international exhibitions. For product development and diversification, the cell collaborates with the National Institute of Design, which investigates the current level of work, production and socio-economic status of the people, interacts with the artisans, and suggests designs and packaging for silk products.

12.3 DEVELOPING MARKETS FOR INSECT PRODUCTS

Production systems and livelihoods are being increasingly influenced by the demands of urban consumers, market intermediaries, and local and international food industries (Van der Meer, 2004). Against this globalized and integrated market backdrop, small-scale farmers, women, indigenous peoples and other vulnerable people face a disadvantage in market participation because they lack access to information, services, technology and credit and the capacity to offer larger volumes of quality products to market agents (Johnson and Berdegué, 2004). Middlemen, oligopolies and monopolies often control markets, thus enabling them to determine substantial price increases and exclude small farmers from participation. In some regions of Mexico, middlemen have been found to exploit local indigenous people, who need money to buy household goods such as pots, clothes and schoolbooks. This often prompts harvesters to continuously exploit the resource, causing pressure on ecosystems and lowering edible insect availability (Ramos Elorduy, Carbajal Valdés and Pino, 2012). Middlemen in Mexico have been found to pay as little as US$30 to collectors for 1 kg of ant larvae of the genus *Liometopum* (known as escamoles), while the national intermediary sells them for US$180 to their international counterparts (Ramos Elorduy, 1997). Setting up insect producers or farmers' associations would help address this problem and would be a crucial step towards developing new markets or diversifying existing markets for insect products.

Although insect gathering and rearing for food and feed in developing countries is carried out informally, selling the insects and/or their products at markets occurs in a more formal way. Insect markets and trade are relatively well structured within their own local contexts and form a network including producers/collectors, middlemen and sellers and processors. Nonetheless, when insects are not seen as an important, or even real, human food source, it can be difficult for actors to enter or create new markets. This section presents the many issues involved in bringing insects and related products to the market and gives examples of the market strategies of some companies that have or are developing specific markets for their insect products.

12.3.1 Accessing markets: the example of street foods

In developing countries, edible insects are mostly marketed as street foods. Insects offered in stalls and local restaurants are particularly popular in southern African and Southeast Asian countries. Street markets have existed for hundreds of years in both rural and urban areas and provide inexpensive food that generally reflects local diets. Street food is mostly sold to consumers directly. It provides an inexpensive food that is ready to eat on a daily basis. As such, consumers choose street food based on cost and convenience. These markets also follow seasonality of production and harvest and allow for variation in consumer diets. The economic contribution of these markets to developing countries is considerable but is often underestimated or neglected (FAO, 2011a).

While it is true that traders may not conduct formal market research, they certainly observe and learn from experience and adapt accordingly. Traders and vendors regularly experiment with and test approaches in marketing and selling: for example, small plates of prepared or unprocessed insects versus bulk sales and plastic bags; larger volumes and packages aimed at other traders and middlemen; and smaller portions aimed at end consumers. This type of testing and adaptation, based on observation and experimentation, is in fact a form of market research, even if not in the formal sense.

FAO published an overview of how this informal market functions that may provide inspiration and guidance for aspiring foodstall operators considering selling edible insects (Box 12.5).

BOX 12.5

FAO Diversification Booklet 18, *Selling Street and Snack Food*

Processing and preparation issues:

- the kinds of instruments and fuel sources needed – e.g. electricity, wood (taking into account cost);
- how to process the food item (taking into account the food item, insect species and traditions);
- how to package and label the food item (taking into account cost);
- how to transport the food item (taking into account cost);
- credit, loans and the availability of microfinancing for starting entrepreneurs.

Continues

Box 12.5 continued

Basic questions:

- What street and snack foods are in demand?
- What prices can be obtained?
- How to ensure food quality and safety?
- Who are the likely competitors?
- What level of sales can be made and where are the best selling locations?
- What are the travelling distances and times from the farm to the selling locations?
- How are the foods items to be sold (e.g. by street hawking or setting up a street stall)?

Marketing strategy:

- The vendor needs to have direct contact with the consumer in order to be proactive.
- Direct feedback of products should be obtained and tasting activities conducted.
- Free gifts and social interactions are important strategies.
- Food quality and safety and observing hygienic conditions are important elements of a marketing strategy.
- The choice of location and display is important, particularly the way in which the products are ordered, but also the impact of the sun on products. The presence of the vendor also constitutes a major part of the display.
- Decide what to sell.

Source: FAO, 2011b.

Packaging material can be made from locally available materials, provided that they are safe and hygienic and do not alter food quality. Leaves are frequently used as wrappers for street foods because they are cheap and readily available. Other locally available materials include clay pots and bowls for yoghurt, wooden boxes for bottles, and sacks made of plant fibres such as jute and cotton.

Street food enterprises are commonly family or one-person businesses and the majority operate without licensing within the informal sector. Studies in developing countries have shown that 20–25 percent of household food expenditure is incurred outside the home, and some segments of the population depend entirely on street foods. Street foods are especially popular in Asia. In Bangkok, 20 000 street food vendors provide city residents with an estimated 40 percent of their overall food intake (FAO, 2011a).

12.4 MARKET STRATEGIES

Insects and their related products can be mass-produced and sold for crop protection (beneficial insects), crop pollination (bumblebees) and health (maggot therapy), as well as for human, pet and livestock nutrition, for research, and a host of other uses on national and international markets, such as collectors' items. Many types of insects are sold live; however, insect products and byproducts probably account for the majority of insect commercialization (Kampmeier and Irwin, 2003).

Markets in developing countries are highly diversified, yet little is known about their establishment and development. In many cases, insects are also exported to other regional markets, as in the case of tarantulas and crickets in Cambodia. Due to the informality of this trade, little accurate information is available regarding the quantity of insects bought and sold on the market (C. Munke, personal communication, 2012).

A wide range of examples of market research and development, as well as strategies by companies to bring insects to markets in developed and developing countries, are given below. These examples show the wide variety of approaches ongoing in different markets in different countries, but are by no means exhaustive.

12.4.1 Insects as exotic foods in the United States

In the 1960s, the North American company Reese Finer Foods started selling chocolate-covered ants, bees, caterpillars and grasshoppers, French fried grasshoppers, French fried silkworms and roasted caterpillars. The initial idea of the Chicago food importer, Max Ries, who founded the company, was to import exotic food items such as Japanese seasoned sliced whalemeat and rattlesnake meat to please exotic palates. The next president of the company developed the idea and imported ants from Colombia and later Japan. The production of this novelty item ceased because of environmental movements against their importation. Today, Reese Finer Foods is still a distributor to grocery stores in the United States, but exotic foods no longer feature in its range of products.

12.4.2 Western approaches today: insects as novelty and exotic foods

The last decade has seen the return of insects to novelty and exotic food stores and delicatessen shops, especially in developed countries. Several kinds of insect have appeared on shelves, or are sold via the internet, in Europe, Japan and the United States. These products range from canned ants and silkworm pupae from Japan to maguey caterpillars from Mexico and fried grasshoppers. Canned white agave caterpillars have been exported to Canada and the United States. These cans contain only five or six larvae per can and sell for US$50 per kg (Ramos Elorduy *et al.*, 2011).

The "novelty" concept is a marketing strategy for selling insects. Fried insects embedded in chocolate or hard candy, and fried and seasoned larvae, can be found in the United States, while the world's most famous luxury stores, Harrods and Selfridges, sell fancy insect products in London. Exclusive chocolates topped with crickets dipped in gold paint are also sold in Brussels. Buying luxury products (with insects) directly from producers via the internet is also possible.

12.4.3 Insects as pet food

Some insects are imported from developing countries to developed countries for sale in pet shops. The Chinese company HaoCheng Mealworm Inc. exports 200 tonnes per year of dried mealworms to North America, Australia, Europe, Japan, Korea, South Africa, Southeast Asia and the United Kingdom of Great Britain and Northern Ireland, among others. The company sells yellow mealworms, superworms and fly maggots. The yellow mealworms are sold alive, dried, canned or processed into a mealworm powder. The superworms are sold alive, dried or canned, and the fly maggots are sold canned (HaoCheng Mealworm Inc., 2012). Mealworms and superworms can be used as a feed supplement for pets, including birds, dogs, cats, frogs, turtles, scorpions and goldfish. According to the company, the mealworms can also be used as human food – incorporated into bread, flour, instant noodles, pastry, biscuits, candy and condiments, and directly into dishes on the dining table.

In the Netherlands, companies that rear insects as pet food now sell mealworms and locusts for human consumption. Kreca is an example of such a company. However, mealworms are still a niche market in the human food industry, and these companies survive mainly through the sale of insects as pet food.

12.5 TRADE

The trade in insects as food to Western countries is driven mainly by demand from migrated communities from Africa and Asia, or by the development of niche markets for exotic foods.

A case study conducted in the Central African Republic observed that the principal importers of caterpillars were Chad, Nigeria and Sudan, via the Economic and Monetary Community of Central Africa. The Central African Republic also exports caterpillars to African communities in Belgium and France (Tabuna, 2000) (see also Box 12.6). Zimbabwe trades caterpillars to Botswana, the Democratic Republic of the Congo,

South Africa and Zambia. Mexican agave worms are exported to the United States (Ramos Elorduy, 2009; Ramos Elorduy *et al.*, 2011). Seven hundred tonnes of edible insects from the Lao People's Democratic Republic and Cambodia are imported for sale in Thailand because of high consumer demand (Yen, Hanboonsong and van Huis, 2013). Edible insects are also exported to the United States for supply to Asian communities (Pemberton, 1988). A particular example of international trade in Asia is the trade in Japanese wasps (Box 12.7).

BOX 12.6

Ethnic foods through migration: the export of caterpillars from Africa to France and Belgium

Mopane caterpillars are mostly exported from Africa to Europe. Annually, Belgium imports 3 tonnes and France 5 tonnes of dried mopane caterpillar (FAO, 2004), mainly from the Democratic Republic of the Congo (Tabuna, 2000). Congolese immigrants from the Congolese quarter in Brussels, Matongé, are the main consumers of the mopane caterpillar.

BOX 12.7

Japanese trade in wasps

Insects are consumed in the mountainous areas of Japan during autumn. Although entomophagy has generally declined, the eating of wasps (*Vespula* spp. and *Vespa* spp.) can still be found. Wasp nests are sold in markets during the harvest season in autumn for the regular price of US$100 per kg. As a result of increased demand, wasps are imported from China, New Zealand and the Republic of Korea. Increases in demand may lead to overexploitation. However, if such insects are to be used sustainably, appropriate commercial use depends on people's awareness of the insects' habitats and their habitat requirements.

Source: Nonaka, 2010.

13. Promoting insects as feed and food

The polarity of views surrounding the practice of entomophagy requires tailor-made communication approaches. In parts of the world where entomophagy is well established, such as the tropics, communication strategies need to promote and preserve edible insects as valuable sources of nutrition in order to counter the growing westernization of diets. In areas where food security is fragile, edible insects need to be promoted as key foods and feeds for nutritional, cultural and economic reasons. However, Western societies still largely averse to the practice of eating insects will require tailored strategies that address the disgust factor and break down common myths surrounding the practice. Governments, ministries of agriculture and even knowledge institutions in developed countries will need to be targeted, given that insects as food and feed are still largely absent from political and research agendas. Insects are still viewed as pests by a large majority of people, despite the increasing literature pointing to their valuable role in the diets of humans and animals.

13.1 THE DISGUST FACTOR

Common prejudice against eating insects is not justified from a nutritional point of view. Insects are not inferior to other protein sources such as fish, chicken and beef. Feelings of disgust in the West towards entomophagy contributes to the common misconception that entomophagy in the developing world is prompted by starvation and is merely a survival mechanism. This is far from the truth. Although it will require considerable convincing to reverse this mentality, it is not an impossible feat (Pliner and Salvy, 2006). Arthropods like lobsters and shrimps, once considered poor-man's food in the West, are now expensive delicacies there. It is hoped that arguments such as the high nutritional value of insects and their low environmental impact, low-risk nature (from a disease standpoint) and palatability may also contribute to a shift in perception (Box 13.1).

BOX 13.1

How can people with an aversion to insects understand and accept that insects are palatable?

Learning to accept insects as food means tackling negative attitudes towards insects in general. A better understanding of *what an insect is* and *what an insect does*, particularly through direct experience, can trigger appreciative reactions, even in the short run (Vernon and Berenbaum, 2004). Further exposure and introduction to entomophagy itself can help to reduce the surprise and novelty of seeing insects on the plate. Zoos, museums and universities can play an important role here. However, the emotion of disgust can be very hard to change.

The question of whether edible insects can be accepted as a food item and become a part of food habits in Western societies depends on at least two crucial factors: availability and learning.

"Bug banquets" (Wood and Looy, 2000; Looy and Wood, 2006) are a combination

Continues

Box 13.1 continued

of educational talk and the chance to directly experience entomophagy. Insects are presented as food that can be tasted, and prejudices are tackled. Years of experimental experience in the Netherlands and the United States have confirmed the effectiveness of bug banquets in overcoming the disgust factor.

Source: F. Dunkel, personal communication, 2012; M. Peters, personal communication, 2012.

Perhaps Bequaert (1921) in his paper, *Insects as Food: How They Have Augmented the Food Supply of Mankind in Early and Recent Times*, best sums up the issues that entomologists are still grappling with:

> In spite of the weight of evidence from an historical point of view, it is not the purpose of the present article to furnish arguments regarding the value of insects as food or for including them in our own diet. **What we eat is, after all, more a matter of custom and fashion than anything else.** ... It can be attributed only to prejudice, that civilized man of today shows such a decided aversion to including any six-legged creatures in his diet. [Emphasis added.]

In general, education is the key instrument for creating public awareness of the potential roles of insects and in influencing consumer choices towards a more balanced and favourable outlook on insects as food and feed; innovative cookbooks can help in this (Box 13.2). Although the disgust factor is more common in Western societies, aversion towards eating insects in the West has also arguably affected people in tropical countries. In Malawi, Morris (2004) found that people living in urban areas and devout Christians reacted with disdain to eating insects. As a result of Western influences, particularly in Africa, research on the contribution of edible insects to nutrition and economy, and on insect species' biology and ecology, has been sporadic (Kenis *et al.*, 2006).

BOX 13.2

Edible insect cookbooks

Chefs and food culture play a large role in determining the acceptance of foods. In some cases, those who would be averse to eating a whole grasshopper might enjoy a mealworm cupcake instead. Here are examples of cookbooks that feature insect recipes:

- *Creepy Crawly Cuisine: The Gourmet Guide to Edible Insects*, by Julieta Ramos Elorduy
- *Eat-a-Bug Cookbook: 33 Ways to Cook Grasshoppers, Ants, Water Bugs, Spiders, Centipedes and their Kin*, by David George Gordon
- *Man Eating Bugs: The Art and Science of Eating Insects*, by Peter Menzel and Faith D'Aluisio
- *Het Insectenkookboek (The Insect Cookbook)*, by Arnold van Huis, Henk van Gurp and Marcel Dicke.

According to UNESCO (2005), the success of education for sustainable development (Box 13.3) hinges on cooperation between all sectors of the education community: formal, non-formal and informal. Taking this as a building block, addressing the entomophagy disgust factor in Western societies might depend largely on the ability to involve the entire educational community. For this reason, engaging all sectors is recommended, particularly in Western societies.

BOX 13.3
Established approaches used in education for sustainable development

Formal education: Primary, secondary, post-secondary and higher education.
Non-formal education: Nature centres, NGOs, public health educators, private companies, private research centres and agricultural extension agents.
Informal education: Traditional and online media, including television, radio, websites, newspapers, magazines, Twitter, blogs, YouTube and Facebook.

Source: UNESCO, 2005.

13.1.1 Edible insects in formal curricula

Until recently, edible insects as a subject – which includes farming techniques, conservation and management issues, as well as insect ecology and biology in the context of food and feed – has been largely absent in formal curricula. Although the topic of insects in biocontrol (e.g. IPM) has been fully entrenched in agricultural sciences for over 35 years (Kogan, 1998), insects in Western sciences are still largely conceptualized as agricultural pests. Therefore, departments of entomology often form part of agriculture faculties rather than science. The past decade, however, has seen a slow but steady rise in food insects in formal education. As of the end of 2011, 46 percent of the 50 land grant universities in the United States – the main food and agricultural universities in the country – had at least one course in their curricula or an annual event that featured food insects (F. Dunkel, personal communication, 2012). Some, such as Montana State University, the University of Illinois and the University of Georgia, have annual events involving 50 to several hundred participants (F. Dunkel, personal communication, 2012). In the Netherlands, the Laboratory of Entomology at Wageningen University offers "insects and society" courses (including entomophagy), which have proved popular among university students. In the Lao People's Democratic Republic, the Faculty of Agriculture at the National University offers "cricket farming" as a subject.

13.1.2 Research and development

Formal education programmes are based on research and are conducted primarily at universities and governmental and non-governmental agencies. This section provides a non-exhaustive overview of ongoing edible insect research and education.

The Netherlands

The Laboratory of Entomology is part of the Plant Sciences Group of Wageningen University. The group carries out fundamental and applied research related to the biology of insects. Its mission is to unravel the ecology of interactions between insects and other ecosystem components by combining ecological studies (population and community levels) with investigations of the underlying mechanisms (subcellular to individual levels). Integrated pest, vector and disease management strategies are being developed in both developed and developing countries. The Chair group has an outstanding reputation in multitrophic interactions, biological control, malaria vector research and entomophagy, and draws worldwide attention to the issue of entomophagy. Professor van Huis coordinates the programme "Sustainable production of insect proteins for human consumption" (known as Supro2) (2010–2013), which is funded by the Ministry of Economic Affairs. The objective is to explore the potential sustainable production of edible insects and insect-derived products, particularly proteins, as a reliable and high-quality food source with a lower negative environmental impact than conventional meat production. Edible insects are reared on organic side streams, after which their proteins are separated and purified and then characterized in order to tailor them for food products.

The laboratory has been instrumental in compiling a list of over 1 900 edible insect species (as of 2012) worldwide. The Netherlands is one of the few countries in the Western world where insect-rearing companies produce insects for human consumption, which are subsequently marketed.

United States

Universities in the United States have been working on promoting insects as food and feed for many years. Montana State University is a leading centre on entomophagy initated by the late Professor Gene DeFoliart. Florence Dunkel is an associated professor of Entomology in the Department of Plant Sciences and Plant Pathology. Her research focuses on plant-based natural products for insect management, particularly related to post-harvest ecosystems worldwide. Current projects include the exploration of Montana wheat varietal resistance to postharvest insects; the use of plant-based products with entomopathogenic fungi for the management of insect pests; and the use of natural products in the holistic management of malaria in West African (Malian) villages. Food insects and an insect feast have formed part of her curriculum in entomology for 24 years, after a tasty introduction to sautéed brown locusts while working in Rwanda. In 1995, Gene DeFoliart invited her to take over the editorship of *The Food Insects Newsletter*, also published as a book (DeFoliart *et al.*, 2009) (Box 13.4).

BOX 13.4
The Food Insects Newsletter[14]

In 1988, several years after Western science had begun to take a strong interest in insects as food, Gene DeFoliart launched *The Food Insects Newsletter*. Funding for some of the early work was provided by forward-looking and interculturally competent programme officers at USAID-Washington. At the time, scientists in the United States had begun to recognize the usefulness of biocontrol and plant-based insect management – techniques of insect management appreciated and understood in native systems for millennia. After a flurry of research it became clear, however, that this research area was not a good fit with the tenure process, either in terms of obtaining funding or attracting graduate students to the research programme. Moreover, public interest and Western scientific attitude were unsupportive. The last 20 years, however, have seen a gradual rise in interest in food insects among the same groups.

[14] Florence Dunkel provided this box.

Denmark

The Faculty of Sciences of the University of Copenhagen specializes in a range of research and teaching on sustainable agriculture, food production and processing, and human welfare related to nature and managed ecosystems. The faculty offers several well-attended international courses at the MSc and PhD levels, including a course on biocontrol. The Department of Plant and Environmental Science also offers courses in sustainable crop production, including the management of pest insects and the protection of beneficial insects – the latter group including honeybees. A research group, Insect Pathology and Biological Control, was established 20 years ago and focuses on insect pathogenic fungi. Today, it is a leading international team in insect pathology and has published many studies on the natural occurrence of insect pathogens among wild and domesticated insects. At the same university, the Research Group on Paediatric and

International Nutrition, which forms part of the Department of Nutrition, Exercise and Sports, developed expertise in performing population-based studies of healthy (in Denmark) and malnourished (in developing countries) infants and children. Chapter 6 contains details of their WinFood project on combating malnutrition with insect-based diets in Cambodia and Kenya. In fact, the present publication contains a variety of nutritional data obtained from studies carried out by Nanna Roos, the coordinator of the WinFood project.

Thailand

Khon Kaen University is the largest public university in northeastern Thailand and is recognized as the regional leader in innovation relating to teaching, learning and research. The Entomology Division falls under the management of the Faculty of Agriculture and carries out teaching and research on useful insects, industrial insects and insect pests to ascertain their effects in agricultural systems, and the management and conservation of insect biodiversity. The Entomology Division pioneered edible insect farming and is one of only three universities in Thailand to conduct research and offer undergraduate and graduate studies in entomology. Professor Yupa Hanboonsong is the contact person and expert in the field, and has conducted several projects concerned with the diversity of edible insects in Thailand, in addition to acting as a technical advisor to the FAO edible insect project in the Lao People's Democratic Republic from 2010 to 2013.

China

The Research Institute of Resource Insects of the Chinese Academy of Forestry, located in Kunming, Yunnan province, is the only national forest research institute in southwestern China. The research institute mainly carries out application-based and basic research related to resources such as insects, economic plants, micro-organisms, vegetation restoration and ecological reconstruction. The research, exploitation and use of resource insects constitute one of the main study objects. Investigated insect species include industrial material insects (such as lac insects, white wax scale insects, gallnut aphids and cochineals), environmental insects, pollination insects, edible and medicinal insects and ornamental insects (butterflies). Research areas include biology, ecology, molecule biology, chemistry, usage, processing of insect materials, mass rearing, artificial cultivation and host plants. The research group, led by Dr Ying Feng, has conducted research for many years, particularly in the southwest, relating to the culture, harvesting and mass-rearing of edible insects in China. They have collected more than 100 specimens of edible and medicinal insects and published more than 20 research papers and two books.

Kenya

In Kenya, Professor Monica Ayieko of Jaramogi Oginga Odinga University of Science and Technology collaborates with other Kenyan research institutes to raise awareness of insects in universities and at the national level. With the limited laboratory resources at her university and by seeking help from other institutions with conventional food analysis technology, she has undertaken basic nutritional analysis of edible winged termites (*Macrotermes* spp.), lake flies (*Chironomus* and *Chaoborus* spp.), the edible grasshopper (*Ruspolia differens*) and the African thief ant (*Carebara vidua*), which are readily available in the Lake Victoria region. In response to requests from consumers, Professor Ayieko has attempted to work on processed products of termites and lake flies and has successfully formulated insect-based biscuits, crackers, muffins, meatloaf and sausages.

The aforementioned WinFood project (facilitated by the University of Copenhagen) collaborated with researchers at the University of Nairobi on termites as a potential additive in baby food.

The International Centre of Insect Physiology and Ecology is a pan-African research and development organization based in Nairobi, Kenya. Its mission is to help alleviate poverty, ensure food security and improve the overall health status of people living in the tropics by developing and extending management tools and strategies for harmful and useful arthropods, while preserving the natural resource base through research and capacity building. The International Centre of Insect Physiology and Ecology will continue to develop, introduce and adapt new tools and strategies for arthropod management that are environmentally safe, affordable, appropriate, socially acceptable and applicable by the target end-users, with full community participation. Its Commercial Insects Programme studies honeybees, stingless bees and silkworms. The centre helps scale up beekeeping and sericulture-based products, as well as pollination services, in several East African countries and the Near East and North Africa region. It also obtains certification and develops market linkages through private entrepreneurs.

Benin

In Cotonou, Benin, the non-profit organization Centre de Recherche pour la Gestion de la Biodiversité (CRGB) has completed numerous studies of an environmental nature, such as fauna and flora inventories, nature conservation and management plans in many French-speaking African countries. Over the years, the centre has accumulated a wealth of experience in entomology, the preservation of cultures, IPM programmes, sustainable agriculture and environmental protection. In 2008, Sévérin Tchibozo, coordinator of CRGB, established a website and database of edible insects, LINCAOCNET, which contains information on edible insects in Benin, Burkina Faso, Cameroon, Central African Republic, the Republic of the Congo, the Democratic Republic of the Congo, Mali, Niger, the Republic of Guinea and Togo. It is the result of collaboration between the CRGB and the Royal Museum for Central Africa of Tervuren, Belgium. The purpose of the LINCAOCNET project is to collect and disseminate information to the people of sub-Saharan Africa about edible insect species and their handling, as well as where to find them and how to catch and prepare them. This information source serves as a basis for better scientific knowledge and for the improved use of insects as food. The project promotes entomophagy by providing information on the management and conservation of edible insects that is accessible to all. The CRGB cooperates with several African and overseas research and development institutions, including the French Global Environment Facility; the Van Tienhoven Foundation; the Netherlands Centre for Biodiversity Naturalis; the French Foundation for Research on Biodiversity; the National Biodiversity Institute in Costa Rica; the International Organization of Francophonie; French Agricultural Research for Development; the Pesticides Initiative Programme of the Europe–Africa–Caribbean–Pacific Liaison Committee; the International Commission for Food Industries; the National Museum of Natural History in Paris, France; and the Royal Museum for Central Africa in Belgium. Benin is part of a recent south–south initiative to exchange traditional and scientific knowledge on food with Bhutan and Costa Rica (Box 13.5).

Mexico

The National Autonomous University of Mexico, the oldest university in Latin America and the best known in Mexico, has an outstanding reputation in edible insect research. Professor Julieta Ramos Elorduy and her team within the Institute of Biology, Faculty of Sciences, are dedicated to the study of biodiversity in the country, employing classical methods as well as molecular biology and electronic microscopy. The Institute has a botanical and zoological department, biological research stations in the east and west of the country, and a botanical garden in Mexico City. The Edible Insects Laboratory is dedicated to the study of edible insects among ethnic groups and the performance of related biological and ecological studies, including on nutritional value. Applied research involves the recognition, identification, collection, preparation, storage, sale and marketing of edible insects. The goal is to improve rural livelihoods and to contribute to regional and national economies.

Professor Ramos Elorduy began to study edible insects in 1974, and in 1982 she wrote a book titled *Insects as a Source of Protein in the Future*. This was followed, in 1984, by *Los Insectos comestibles en el México antiguo* (Edible Insects in Ancient Mexico) and in 1998 by *Creepy Crawly Cuisine*. She founded the Ethnobiological Scientific Society and organized the First Congress in Ethnobiology in 1994. She is recognized as an expert in edible and medicinal insects.

Lao People's Democratic Republic

In the Lao People's Democratic Republic, the Faculty of Agriculture at the National University of Laos has initiated an innovative and successful programme to introduce insect farming to students, coupled with awareness-raising on the nutritional value of insects and livelihood potential. Students are taught basic insect-farming practices and raise crickets in groups, culminating in harvesting and insect eating at a major social event. Some students have subsequently introduced insect farming to their families (P. Durst, personal communication, 2012).

13.2 DRAWING ON TRADITIONAL KNOWLEDGE

13.2.1 Insect farmers and collectors

The primary producers of edible insects are farmers and collectors. In most cases, indigenous knowledge forms the basis of sustainable collection and harvest practices. Therefore, it is important to document and promote sustainable traditional best practices to share with others. To this end, education on best practices, training and the creation of associations to enable knowledge-sharing can help both farmers and collectors.

Several individuals, organizations and companies have raised questions on good farming practices, markets, and processing and legal requirements, in particular regarding regulations on the use of insects in food and feed products. Since this demand exists, governments may wish to expand their technical capacities within their agricultural (extension) services. An example of possible support is the FAO-assisted Technical Cooperation Programme in the Lao People's Democratic Republic, "Sustainable insect farming and harvesting for better nutrition, improved food security, and household income generation".

An interesting aspect in training programmes is the merging of traditional knowledge with new technologies. In Kenya, indigenous ways of collecting termites (*Macrotermes subhyalinus*) have been improved to ensure reliable collection; for example, a new light trap was designed in collaboration with the Kenya Industrial Research and Development Institute (Ayieko *et al.*, 2011).

13.2.2 Cultural and gastronomic activities (festivals, expo, restaurant, museums)

Cultural and gastronomic activities include festivals, art and scientific exhibitions in museums and zoos, insect menus in restaurants and snacks in bars, and cooking workshops. Governments, knowledge institutions, farmers and producers, among others, can sponsor such activities.

Museums

In 2008, the Royal Museum for Central Africa in Tervuren, Belgium, initiated a project called the Central African Biodiversity Information Network (CABIN). The aim of this project – funded for five years by the Belgian Cooperation and Development Agency – is the implementation of a network of databases on biodiversity, in collaboration with several research institutions in Central Africa (mainly Burundi, the Democratic Republic of the Congo and Rwanda). The LINCAOCNET database project (see section 13.1) was established by CRBG in the context of the CABIN project.

The Natural History Museum in London, home to one of the world's richest entomological collections, has also shown interest in the topic, setting up a travelling exhibition in London shopping centres on the theme of edible insects (Fairman, 2010). Additionally, the Victoria Bug Zoo, located in British Columbia, Canada, offers visitors hands-on interaction with insects.

BOX 13.5

International knowledge-sharing between developing countries on the use of edible insects in diets

A programme for the exchange of traditional and scientific knowledge on food has been launched by Benin, Bhutan and Costa Rica. Experts from the Faculty of Agronomic Sciences of the University of Abomey/Calavi, in Benin, the National Mushroom Centre in Bhutan and the National Institute of Biodiversity in Costa Rica have joined forces in this initiative.

In particular, Costa Rica and Bhutan are obtaining information from Benin regarding which insects are edible and can be introduced to people's daily diets. Costa Rica is sharing knowledge on how to classify the various species of insect and the different ways in which they can be used, and on how to breed and preserve them efficiently. Although such insects constitute an important part of the diet of people in Benin, expertise on how to best exploit them is lacking.

In the meantime, the National Institute of Biodiversity is working to change the attitude of people in Costa Rica towards insects. The country is host to 365 species of insect, many of which could be used as feed for farm animals.

Source: Cooperation, 2012.

13.2.3 Recent examples of key results

Research on edible insects has concentrated on the traditional food habits of indigenous people. Julietta Ramos Elorduy has published an impressive number of articles about entomophagy in Mexico from 1977 up to the present (see section 13.1). A landmark was the publication in 2005 of *Ecological Implications of Mini-livestock: Potential of Insects, Rodents, Frogs, and Snails*, edited by Professor Maurizio G. Paoletti at the University of Padua in Italy. The book contains contributions by many authors covering various aspects of entomophagy from around the world.

There have been three international meetings in which edible insects figured prominently:

- In 2000, the conference *Les Insectes dans la tradition orale* [Insects in oral literature and traditions] took place in Paris. It was ethnologically oriented and had edible insects as a topic. The proceedings were published in 2003 (Motte-Florac and Thomas, 2003).
- In February 2008, FAO organized a workshop in Chiang Mai, Thailand, titled Forest Insects as Food: Humans Bite Back. It focused on edible insects in the Asia-Pacific region.
- In January 2012, FAO and Wageningen University jointly organized the Expert Consultation Meeting on Assessing the Potential of Insects as Food and Feed in Assuring Food Security at FAO in Rome. This consultation was the first of its kind co-organized by FAO (see section 1.2).

The topic of insects as a source of food and feed has attracted strong media attention in the last couple of years: international and national newspapers, TV stations and other media sources have produced articles and documentaries on the topic. Strong media attention is helping to influence public policymaking regarding, for example, the review of food and feed regulations with respect to the use of insects.

13.3 ROLE OF STAKEHOLDERS

The communication strategies of stakeholders need to be comprehensive and should target region, culture, locality (rural, urban), economics, environment, nutrition, gastronomy and tradition. In developing countries, different approaches are needed for urban, peri-urban and rural communities.

13.3.1 Governmental bodies

Governmental bodies have important roles to play in promoting insects as food and feed. In particular, the development of this new sector as a viable (and environmentally friendly) alternative to the conventional food and feed sector will require that governmental bodies address the following issues:

- awareness and collaboration among relevant ministries, such as agriculture, health and the environment;
- the implementation of existing policies and the creation of new policies, such as food and feed regulations;
- the creation of incentives aimed at knowledge centres for research, development and graduate and post-graduate training;
- the creation of incentives aimed at the private sector for investment and technical development;
- the provision of technical assistance in sustainable insect harvesting and insect farming through agricultural extension services.

One example of government incentives is the three-year, €3 million EU FP7 project "Insects as novel sources of proteins". Starting in the second half of 2013, this collective research project will involve various universities and companies in examining ways of rearing and processing insects for feed.

An effective communication strategy needs to differentiate between insects as food and as feed and also to minimize sensationalism about insect consumption by using well-documented literature to increase credibility. Among the strategies to be considered for developing effective communication strategies for governments, international agencies, the private sector and NGOs are: tailoring messages for different audiences; identifying incentives for using insects as food; using success stories and best practices/experiences to promote the consumption of insects; involving (local) media to raise awareness; creating a communication toolkit on the importance of and opportunities for insects as food and feed; and seeking endorsements from celebrities to improve the credibility of the sector.

13.3.2 Industry

Industrial producers have undertaken research and development on insects in cooperation with knowledge institutions with the objectives of centralizing scattered information including data, literature, economics, methods and practices as a basis for investment options. Industry can further advance insects on the agenda by contributing to investment in infrastructure, research and technology, and can increase awareness by marketing products to the general public.

The industry also has good contacts with regulators and policymakers. It could take a proactive stance by facilitating the development of regulations with government agencies.

The industry could also develop a roadmap on insect protein technology for the private sector. At the Expert Consultation Meeting in January 2012, stakeholders from the private sector emphasized the need to create an international industry association to support insect-sector initiatives. These could include effective awareness-raising with the general public and the use of a common language by industrial stakeholders to avoid confusion and help ensure effective marketing.

13.3.3 Non-governmental organizations

NGOs play a significant role in increasing awareness of entomophagy, as well as in promoting insect rearing as a diversified livelihood strategy. Environmentally oriented NGOs can help to strengthen guidelines for sustainable harvesting through governmental lobbying and practical experience in local communities. NGOs can also raise awareness of this already significant informal activity and promote it as an environmental strategy for food and feed on political agendas in both developed and developing countries.

Moreover, NGOs can assist in technical training for rural, peri-urban and urban households on market linkages, entrepreneurship, the domestic rearing of insects, and the identification of producers' objectives (such as subsistence, semi-commercial and commercial enterprises). Examples of such projects are the Insect Centre in the Netherlands and BugsforLife in Benin.

Other resources available online include the Bay Area Bug Eating Society and the popular website Girl Meets Bug, maintained by Daniella Martin.

13.3.4 Gastronomic enterprises

Making insects tasty and attractive is one of the biggest challenges facing new insect-based food enterprises. Initiatives like the Nordic Food Lab in Copenhagen (Box 13.6) and the Ento project in London are examples of this larger effort for palatability. These organizations focus on optimizing colour, texture, taste and flavour to make insects appealing to the Western palate. The Tokyo Mushikui (bug-eating) Festival is attempting to revive interest in edible insects in Japan (Box 13.7).

BOX 13.6

The Nordic Food Lab

How do you take something considered inedible, like an insect, and have it recategorized as edible? One of the many powers of cooking, and science in general, is that it can bring to us a new understanding and appreciation of the world. Instead of serving a cricket whole on a plate, as other attempts at normalizing entomophagy have done, in this case it is more effective to transform the raw material into something that will be recognized as delicious before edibility is even raised as an issue. If it looks, smells and tastes delicious, it must be edible.

The strategy of the Nordic Food Lab is based on the following assumption: instead of accepting, as contemporary culture does, that something must be edible before it can be delicious, these two categories are viewed as distinct, though overlapping, like a Venn diagram. Just as there are foods that are edible but not necessarily delicious (certain "weeds", for example), there are foods that are delicious before edibility is considered in popular consciousness. It is this boundary that the Nordic Food Lab wants to push – to explore the vast range of delicious flavours in order to incorporate an increasingly wide array of foods into the sphere of the edible.

The Nordic Food Lab is a non-profit organization that explores the building blocks of Nordic cuisine through traditional and modern gastronomies, generating knowledge for chefs, industry and the public. Much of its research has focused on wild foods like plants, seaweed, shellfish, game and edible insects.

Source: Nordic Food Lab, 2012.

BOX 13.7
Konchu Ryori Kenkyukai

In Japan, the Konchu Ryori Kenkyukai (Insect Cuisine Research Association) was created to acknowledge the presence of insect delicacies in historical Japanese cuisine. The group organizes the Tokyo Mushikui (bug-eating) Festival, which celebrated its fourth year in 2012. The group's first bug-eating festival drew only 30 participants, but numbers have more than doubled since.

There are many traditional insect dishes in traditional Japanese cuisine. For example, sanagi, a silkworm delicacy, used to be a fairly common dish and is still available canned. Other formerly common delicacies were inago (grasshoppers, often steeped in sugar and soy sauce), hachinoko (bee larvae) and zazamushi (stonefly larvae). Many Japanese today, however, have never tried such dishes. One reason behind the Tokyo Mushikui Festival is to revive this older culinary culture, as well as to stoke interest in new flavours.

Source: adapted from Tempelado, 2012.

In much of its work, the Nordic Food Lab explores the relationship between edibility and deliciousness, asking questions like, What makes something good to eat and why? How can we come to understand more deeply the systems that tell us, in different places and times? What can we – and what should we – put "inside" to transform the "outside" into part of us? By exploring the vast range of flavours, the Nordic Food Lab aims to turn "inedibles" into edible ingredients. Seaweed is one such food source: just a few years ago it was considered in the West as either exotic or niche, but now, in certain places, it is celebrated as a new, versatile ingredient – since it was shown to be delicious (Nordic Food Lab, 2012). The head of the culinary research and development group says that deliciousness is the first and most important factor in developing new gastronomic building blocks. Mayonnaise from bee larvae works not because of its novelty but rather because of its earthier and more satisfying taste – its unique deliciousness (Baines, 2012).

The Ento project is a roadmap for introducing edible insects to the Western diet. This group of designers of the Royal College of Art and Imperial College in London has tackled the issue of sustainability from an innovative, design-driven approach. Ento focuses on acceptance and proposes the creation of a culture surrounding insects. Ento took sushi as an example of a recently accepted food and used it as the inspiration for its design concept. They created a roadmap for launching insects as a new food, focusing on different groups of the public at different stages. The underlying logic is that not everyone will suddenly start to eat new foods and therefore it is necessary to target the more adventurous user before eventually offering it in supermarkets as a normal, everyday food item.

Ento conducted tasting tests of various processed insects and concluded that the power of abstraction was critical for food design as well as for the entire branding of the company. Their "entocubes" abstract the animal behind the food and emphasize the cleanliness, human control and futuristic aspect of insects as food. Ento conducts taste experiments using different types of processing, such as boiling, frying and baking. Based on a technique called molecular food pairing, they have created a database of foods that could be used with insects to create new recipes (Ento, 2012).

In San Francisco, curious market-goers wait in line to get wax moth larvae tacos and mealworm ice cream at Monica Martinez's company, the Don Bugito Prehispanic Snackeria. Her company revives prehispanic traditional Mexican foods, selected because they are nutritious and sustainable (see section 6.3).

At an industrial scale, the meat industry can be used as a model for experimenting with insect processing for product development. Just as processed meat contains non-meat ingredients, insect products may contain non-insect ingredients. In addition, several physical and chemical properties of insect products need to be taken into account, such as structure, pH changes, colour, water-holding capacity and flavour.

14. Regulatory frameworks governing the use of insects for food security

The production, trade and use of edible insects as food and feed touch on a wide range of regulatory areas, from product quality assurance to the environmental impact of insect farming. Regulatory frameworks referred to in this chapter include legislation, standards and other regulatory instruments (legally binding or otherwise), at both the national and international levels, which would have a role in regulating the use of insects as food and feed. Regulatory frameworks on the use and conservation of insects in areas such as biodiversity conservation, disease control, IPM, sanitation, pest eradication and the health sector are not discussed here.

Globalization and growing consumer concern over food quality and production methods have dramatically changed consumption patterns in recent decades. Food chains have become longer and more complex due to the global trade in raw materials and food ingredients. As a result, food safety and the quality of traded food products have received increased attention and the regulatory frameworks governing food and feed have developed greatly in the last 20 years.

In many societies, insects are not perceived as a regular food/feed product and, as such, they rarely fall within the remit of food/feed regulators. At the national and international levels, standards and regulations acknowledging the use of insects as ingredients for food and feed are rare (Box 14.1).

BOX 14.1

FAOLEX

FAOLEX is a comprehensive and up-to-date computerized legislative database that constitutes one of the world's largest electronic collections of national laws and regulations on food, agriculture and renewable natural resources. The resource is free and available online (http://faolex.fao.org). A search (conducted on 29 January 2013) in FAOLEX using the keyword "insects" generated 937 references to laws (in over 50 countries) dealing with insect-related issues, mainly concerning sanitation and pest control in the agricultural sector. Laws and regulations on beekeeping and silkworm-raising are well developed in several countries with significant honey and silk industries. A few countries have legislation that refers to insects in food as impurities and that prescribes maximum permissible levels. No reference could be found in the database to countries with laws or regulations governing the use of insects as food or feed ingredients. It seems, therefore, that specific legislation to regulate the use of insects in food and feed is yet to be developed.

At most, legislative references to insects in the context of food prescribe maximum limits of insect traces in foodstuffs, where this is unavoidable. Examples can be found in regulations governing the production of dried products such as grains, flour, peanut butter, fruits, spices and chocolate.

The absence of specific legislation is not because the risks are being neglected but because the quantities of insects in food and feed are, at present, negligible. If insects were to become a more widely used ingredient in food and feed, a risk assessment would need to be carried out and an appropriate regulatory framework created.

For example, according to the United States Food and Drug Administration's (FDA) booklet, *Food Defect Action Levels*, average contamination levels below 150 insect fragments per 100 g of wheat flour pose no inherent health hazard. Table 14.1 gives other examples from the booklet of maximum permissible levels of insect contamination in food products for humans (below which such contamination is not considered to be hazardous to health).

TABLE 14.1
Maximum permissible levels of insect contamination in food products

Product	Type of insect contamination	Maximum permissible level
Canned sweet corn	Insect larvae (corn ear worms or corn borers)	Two or more 3 mm or longer larvae, cast skins, larval or cast skin fragments, the aggregate length of insects or insect parts exceeds 12 mm in 24 pounds
Canned citrus fruit juices	Insects and insect eggs	Five or more Drosophila and other fly eggs per 250 ml or 1 or more maggots per 250 ml
Frozen broccoli	Insects and mites	Average of 60 or more aphids and/or thrips and/or mites per 100 grams
Hops	Insects	Average of more than 2 500 aphids per 10 grams
Ground thyme	Insect filth	Average of 925 or more insect fragments per 10 grams
Ground nutmeg	Insect filth	Average of 100 or more insect fragments per 10 grams

Source: USFDA, 2011.

Insect "impurities" may actually be good for health. For example, people in rice-eating regions typically ingest significant numbers of rice weevil (*Sitophilus oryzae*) larvae, and this has been suggested as an important source of vitamins (Taylor, 1975).

For developed countries, the absence of specific legislation on the use of insects as food and feed ingredients is due to, among other factors, the very limited development of industrial insect farming to supply the food and feed sector and the insignificant quantities of insects currently consumed as food.

14.1 MAJOR BARRIERS FACED

Investors, farmers and entrepreneurs willing to build up industrial-scale insect-rearing plants for food and feed have difficulty identifying the appropriate regulations and laws, if they exist at all. In many countries, the lack of a legal framework on insects for use as food and feed is considered by investors to be a major barrier (Box 14.2).

According to some producers of insects for food and feed, the barriers to establishing markets for insects and the implications for trade are a result of the following factors:

- Unclear regulations and legislation on farming and selling insects for human consumption and feed are an obstacle. For example, in the United States, the FDA's *Food Defect Action Levels* lists allowable percentages of insect fragments in food, yet insects as food do not seem to fall into any category. In the EU, the European Novel Food Regulation (Regulation (EC) No.258/97) (European Commission, 1997), which regulates food and ingredients that were not used for human consumption to a significant degree prior to 15 May 1997, restricts the trade of insects, even if they are consumed in other countries (Lähteenmäki-Uutela, 2007).
- Difficulty in understanding relevant national and international information regarding processing and quality, little networking among producers, and a lack of demand for large quantities for human consumption in developed countries represent additional obstacles.
- The lack of awareness among consumers and buyers about existing markets leads to low demand.
- It is difficult to market insects for human consumption because they are perceived to be inherently unsanitary.

BOX 14.2

Barriers to market establishment in the European Union

Perceived major barriers to the farming of insects in the EU include:

- strict sanitary regulations for setting up farms;
- a lack of guidelines on the mass-rearing of insects;
- a lack of clarity on which insects are to be authorized for the market by EU Novel Food;
- limited information on species eaten prior to 15 May 1997, which is required for categorization as a Novel Food (Box 14.3);
- the recent restriction in the EU on the feeding of poultry, pig and farmed fish with processed animal proteins (PAPs), with no reference to insects.

Source: L. Giroud, personal communication, 2012.

Key EU legislation on the feeding of animals with insect meal can be summarized as follows:

- The catalogue of feed materials (Commission Regulation (EC) No.68/2013) is a non-exhaustive list. In principle, therefore, non-listed products can also be placed on the market. Feed producers are encouraged to have the important feed materials listed and thus described. Although "terrestrial invertebrates" are listed under entry 9.16.1 ("Whole or parts of terrestrial invertebrates, in all their life stages, other than species pathogenic to humans and animals; with or without treatment such as fresh, frozen, dried"), a specific entry for "insect meal", for example, does not yet exist. Such a listing could be made at the initiative of a stakeholder via the EU Feed Chain Task Force.
- According to Regulation (EC) No. 1069/2009, insect meal is a PAP that has to be processed in line with the standards contained therein. Regulation (EC) No. 1069/2009 classifies insects and other invertebrates as Category 3 material (fit but not intended for human food chains). As such, they are suitable as feed for livestock, especially fish, poultry and pigs.

However, despite Regulation (EC) No. 1069/2009, Regulation (EC) No. 999/2001 prohibits the feeding of farmed animals with PAPs, with the exception of hydrolysed proteins;[15] proteins derived from insects fall within the definition of PAPs.[16] Thus, insect meal cannot currently be used as feed in the EU for food-producing animals, and may only be fed to pets. Referring to Regulation (EC) No. 999/2001 (the "BSE" regulation), authorities at various levels support a ban on feeding insect protein to farmed animals. However, in its original version, the BSE regulation only contained a ban on the use of protein from mammals as feed. This is still evident in the current version in the preamble and Article 7(4).

In July 2012, a relaxation of this ban was agreed to allow such PAPs to be fed to aquaculture species. This change will formally commence in early 2013 and will apply from 1 June 2013. Once certain conditions can be met, the EU also intends to re-authorize the use of such PAPs in pig and poultry feed (W. Trunk, personal communication, 2012).

The EU promotes free-range farming of pigs and poultry, in which the consumption of invertebrates is not merely tolerated but seen as correct procedure in terms of both animal welfare and feed intake, as free-roaming poultry and pigs eat insects as a natural feed. However, "natural feed" is not subject to checks for pollutants such as heavy metals, PCBs/dioxins or pesticides.

---

[15] Regulation (EC) No. 999/2001, Annex IV (as amended by Commission Regulation (EC) No. 1292/2005); Regulation (EC) No. 1923/2006; and Commission Regulation (EU) No. 56/2013).

[16] Commission Regulation (EU) No. 142/2011, Annex I(5) defines processed animal proteins as "animal protein derived entirely from Category 3 material, which have been treated ... so as to render them suitable for direct use as feed material or for any other use in feedingstuffs, including petfood, or for use in organic fertilisers or soil improvers".

14.2 LEGAL FRAMEWORK AND STANDARDIZATION

International standards can serve as a useful basis for countries to set up their regulatory frameworks for food and feed. Aligning legislation with international standards, particularly Codex Standards (Box 14.3), facilitates their compliance with trade rules and enables and facilitates the trade of food and feed products.

BOX 14.3

Codex Alimentarius

As the international reference standards for food and feed, a Codex Alimentarius standard on the use of insects as food and feed ingredients could serve as a reference for national legislation on insect production and use as food and feed, from both safety and quality viewpoints.

While the Codex Alimentarius does not contain specific standards on fresh or processed insects for use as food and feed, "insects" are included in the Codex Alimentarius standards as "impurities". For example, Codex Standard 152-1985 stipulates that wheat flour shall be free of:

- abnormal flavours, odours and living insects;
- filth (impurities of animal origin, including dead insects) in amounts that may represent a health hazard to human health.

FAO and WHO established the Codex Alimentarius Commission in 1963 with the main objectives of protecting the health of consumers and ensuring fair trade practices in the food trade.[17] Presently, the Commission has 185 members – 184 member countries, 1 member organization (the EU) and 204 observers. The Codex Alimentarius Commission develops harmonized international food standards, guidelines and codes of practice that contribute to the safety, quality and fairness of the international food trade. Codex standards are based on the best available science assisted by independent international risk assessment bodies and ad-hoc consultations organized by FAO and WHO. Although their implementation by members is voluntary, Codex standards serve in many cases as a basis for national legislation.

Under the World Trade Organization's (WTO) Agreement on the Application of Sanitary and Phytosanitary Measures (SPS Agreement), national legislation that conforms with Codex standards is presumed to comply with international obligations resulting from the SPS Agreement. According to the SPS Agreement:

> Members shall base their sanitary or phytosanitary measures on international standards, guidelines or recommendations ... Sanitary or phytosanitary measures which conform to international standards, guidelines or recommendations shall be ... presumed to be consistent with the relevant provisions of this Agreement and of GATT 1994 (Article 3, paragraphs 1 and 2).

These "international standards, guidelines or recommendations" for food safety are defined, in turn, as:

> the standards, guidelines and recommendations established by the Codex Alimentarius Commission relating to food additives, veterinary drug and pesticide residues, contaminants, methods of analysis and sampling, and codes and guidelines of hygienic practice (SPS Agreement, Annex A, Paragraph 3(a)).

WTO members that wish to apply stricter food-safety measures than those set by Codex may be required to justify these measures scientifically (WHO/FAO, 2012).

---

[17] See the homepage of Codex Alimentarius at: www.codexalimentarius.org.

Specific legal provisions on the use of insects for food and feed production would serve to control and regulate the use of insects by industry processors and would guarantee consumer access to information. For this purpose, regulators would need to assess the potential risks associated with the use of insects, in terms of both species and quantities. Legal frameworks that are more protective of consumer interests may also focus on the information displayed on food packaging and the information made available to consumers on the results of risk assessments on the effects of insects on human heath.

In 2010, the Government of the Lao People's Democratic Republic proposed to the FAO/WHO Codex Coordination Committee for Asia that standards for regional trade and food safety for house crickets be developed. This proposal was not accepted, however, as data indicated that there was no verifiable level of trade in insects to warrant such action (FAO, 2010a).

There is an increasing lobby in the feed sector for the development of specific legislation on the use of insects as feed. Lobbying is being conducted at a national level (including private sector-led actions by United States-based companies to obtain FDA approval on the use of insects in feed) and at a European level.

The farming of insects and edible arthropods for human food, referred to as minilivestock, is emerging as an ecologically sound form of animal husbandry. Recently, as a result of strong lobbying by the feed sector, initiatives have begun to emerge to create an enabling environment for the development of regulations and standards on the use of insects for aquaculture feed and to a lesser degree as human food. At the European level, for example, the quality and safety criteria for insect-based feed are currently undergoing review.

The production and consumption of insects should also be analysed from the viewpoint of their potential impact on health and biodiversity and the potential environmental hazards associated with insect production and release, including the accidental release of insect species not indigenous to the area of production. Risk assessments and containment measures should address potential outbreaks of disease that may be harmful to human or animal health and to plant protection. Other target areas for legislation could include regulation of the trade between countries of *living* insects as breeding stock.

As is often the case in an emerging industry, and as identified by the Expert Consultation Meeting on Assessing the Potential of Insects as Food and Feed in Assuring Food Security in Rome in January 2012, in addition to the development of public national and international standards, self-regulation by the insect producers/processors and other stakeholders in the sector can be useful in fostering harmonization and mutual recognition. This could include the development of standards, codes of practices/ standards and product quality metrics to garner credibility.

The concept of novel food is guiding the development of rules and standards for insects as human food. The term novel food refers to food products that do not have a history of human consumption in the region or country in question. Examples of definitions of novel foods contained in national legislation are:

- "a food that does not have a history of human consumption in Australia or New Zealand" (Australia New Zealand Food Standards Code – Standard 1.5.1);
- "a substance, including a microorganism, that does not have a history of safe use as food" (Canada's Food and Drug Regulations (C.R.C., c.870) – B.28.001).

The term may include edible insects, oils, berries and foodstuffs that are the product of biotechnology (including genetically modified foods). Foods that are the product of biotechnology may be considered as novel globally, but foodstuffs derived from natural products, while novel in some countries, may represent a considerable portion of normal dietary intake in other countries. It has been suggested that a "long history of human use" implies that insects intentionally harvested for human use or consumption do not pose a significant risk (Banjo, Lawal and Songonuga, 2006b). However, in a number of developed countries, such as the United States, Canada and in the EU, some edible insect species may require a premarket safety evaluation and authorization for sale as a novel food or ingredient (Box 14.3).

BOX 14.4
Definition of novel food by the European Commission

Regulation (EC) No. 258/97, Article 3(1), of the European Parliament and of the Council of 27 January 1997 concerning novel foods and novel food ingredients considers foods and food ingredients that have not been used for human consumption to a significant degree in the EU before 15 May 1997 to be "novel foods" and "novel food ingredients".

According to this regulation, such novel foods and novel food ingredients must be, among other things (EC, 1997):

- safe for consumers; and
- properly labelled to not mislead consumers.

The EU obligates risk assessments (premarket safety assessments) on each individual species/product before it is placed on the market and can, in a second step, conduct an authorization (conditions of use, labeling and designation).

The European Food Safety Agency (EFSA), whose function is to ensure consumer protection and maintain food safety in Europe, carries out risk assessments for prospective novel foods. In 2011, the EFSA began to identify media agencies reporting on edible insects as a subject of the primary filter of the agency, which carries out data analysis and signal detection. During this process, EFSA started to collect more information on the subject. The EU is funding a project to investigate insects as a potential protein source in feed (KBBE 2012.2.3-05, see Chapter 13) and is also involved in a network of stakeholders in the field. To include insects in the diet as novel food, a premarket safety evaluation will need to be set up and the EFSA will be asked to perform a risk assessment.

The novel food concept may impose heavy administrative burdens and costs. Therefore, while it may protect consumer health, it may be out of reach of small-scale farmers wishing to farm a "novel" insect species. In the EU, a current proposal on the use of insects for human foods is to consider all insects as novel foods, except those 5–10 insect species (still to be defined) that were most commonly reared in Europe before 1997.

In a given country, a case could be made that an edible insect species is not novel and has a history of safe use. Most probably, the regulatory authority would need to conduct a risk assessment, and additional information such as pathogen/insect combinations would need to be available.

Much work still needs to be done and many issues taken into account when elaborating normative frameworks and adjusting for insect-inclusive food laws. At the Expert Consultation Meeting in January 2012, therefore, a working group developed the following proposals for elaborating regulatory frameworks (FAO, 2012f):

- Scientists, industry and regulators need to collaborate proactively and contribute to self-regulation in the sector. An analysis of existing policies and regulations on food and feed ingredients is necessary and can be achieved by:
  - communicating with the relevant regulatory bodies and their key contact persons;
  - identifying impediments and finding out where the existing framework needs to be improved.
- The development of new policies is inevitable. It will be necessary to listen to regulators to find out what can be expected, to be sensitive to consumers who might demand specific regulations, and to collaborate with retailers. An interesting example of a model to facilitate such a consultation is the Global Agricultural

Practice initiative. Examples of new regulations to be considered include quality standards, quality control and quality assurance guidelines on contaminants and nutritional compositions; label requirements; environmental impact assessments; and requirements for feed for animals (for example, can manure be used?).
- Public-sector and private-sector regulatory frameworks will need to be standardized at the national and international levels.

A defined degree of guaranteed safety is necessary for any product. Feasible practices for production (including sanitation measures) need to be developed, for which other industries can serve as role models. Private and public standards may establish the basis of harmonized regulatory practices on the use of insects as food and feed. Legal frameworks should be developed to consolidate and set binding provisions and to ensure the implementation and enforcement of such provisions throughout the sector. International harmonization through private or public standards would be positive for the sector but difficult to achieve because of the diversity of insect species and processing methods. Nevertheless, this dilemma needs to be addressed.

Recommendations for the regulatory framework on edible insects are as follows:
- Promote private and public standardization at the national and international levels for insects as food and feed, accompanied by a premarket safety evaluation (under Codex Alimentarius, among other standard-setting organizations).
- Promote the establishment of appropriate international and national standards and legal frameworks to facilitate the use of insects as food and feed and the development and formalization of the sector.
- Take into consideration the potential effects of insect production and rearing on the environment, and the environmental and trade implications of the international movement of insects, when drafting and implementing regulatory frameworks for insect production and use. This would oblige regulators to pay attention to a broad range of regulatory areas, including phytosanitary legislation, biodiversity, disease control and the environment.

15. The way forward

Recent developments in research and development show edible insects to be a promising alternative for the conventional production of meat, either for direct human consumption or for indirect use as feedstock. Nevertheless, a tremendous amount of work still needs to be done by a wide range of stakeholders over many years to fully realize the potential that insects offer for food and feed security. The roadmap drawn up during the Expert Consultation Meeting on Assessing the Potential of Insects as Food and Feed in Assuring Food Security in Rome in January 2012 summarized the main tasks that lie ahead:

- Further document the nutritional values of insects in order to promote insects more efficiently as a healthy food source.
- Investigate the sustainability and quantify the environmental impacts of harvesting and farming insects compared with traditional farming and livestock-raising practices.
- Clarify and augment the socio-economic benefits that insect gathering and farming can offer, with a focus on improving the food security of the poorest of society.
- Develop a clear and comprehensive legal framework at the (inter-)national level that can pave the way for more investment, leading towards the full development (from the household scale to the industrial scale) of production and trade in insect products for food and feed internationally.

The case needs to be made to consumers that eating insects is not only good for their health, it is good for the planet. Additionally, insect rearing should be promoted and encouraged as a socially inclusive activity. Rearing insects requires minimal technical knowledge and capital investment and, since it does not require access to or ownership of land, lies within the reach of even the poorest and most vulnerable members of society. In the future, as the prices of conventional animal proteins increase, insects may well become a cheaper source of protein than conventionally produced meat and ocean-caught fish. For this to occur, there will need to be significant technological innovation, changes in consumer preferences, insect-encompassing food and feed legislation, and more sustainable food production.

Insects can contribute to food security and be a part of the solution to protein shortages, given their high nutritional value, low emissions of GHGs, low requirements for land and the high efficiency at which they can convert feed into food. The production of insect biomass as feedstock for animals and fish can be combined with the biodegradation of manure and the composting and sanitizing of waste. Insects can partly replace the increasingly expensive protein ingredients of compound feeds in the livestock, poultry and aquaculture industries. Grains now used as livestock feed, which often comprise half the cost of meat production, could then be used for human consumption (van Huis, 2013).

Considering that insects already form part of the human diet in many countries, their potential needs to be re-evaluated. The sustainable harvesting of edible insects in the wild requires nature conservation strategies. Habitat manipulation measures can increase the abundance and accessibility of insect populations. The possibility of simultaneously controlling pest insects by harvesting them as food/feed should be exploited. Simple rearing procedures for some promising insect species need to be developed. Micronutrient bio-availability (particularly of iron and zinc) in edible insects needs further investigation, given the massive occurrence of these deficiencies in the tropics.

In the Western world, consumer acceptability will be determined, in large part, by pricing, perceived environmental benefits, and the development by the catering industry of tasty insect-derived protein products. Preservation and processing techniques are needed to increase shelf life, conserve quality and increase the acceptability of insect food products; processing procedures are also needed to transform insects into protein meal for animal/fish feedstock and for the extraction of insect proteins to be used as ingredients in the food industry.

Considering the immense quantities of insect biomass needed to replace current protein-rich ingredients such as meal and oil from fish and soybeans, automated mass-rearing facilities that produce stable, reliable and safe products need to be developed. The challenge for this new industry will be to ensure the cost-effective, reliable production of an insect biomass of high and consistent quality. Regulatory frameworks need to be developed. The close collaboration of government, industry and academia will be essential for success.

References

Aarnink, A.J.A., Keen, A., Metz, J.H.M., Speelman, L. & Verstegen, M.W.A. 1995. Ammonia emission patterns during the growing periods of pigs housed on partially slatted floors. *Journal of Agricultural Engineering Research*, 62(2): 105–116.

Acuña, A.M., Caso, L., Aliphat, M.M. & Vergara, C.H. 2011. Edible insects as part of the traditional food system of the Popoloca town of Los Reyes Metzontla, Mexico. *Journal of Ethnobiology*, 31(1): 150–169.

Adamolekun, B. 1993. *Anaphe venata* entomophagy and seasonal ataxic syndrome in southwest Nigeria. *Lancet*, 341(8845): 629.

Adamolekun, B., McCandless, D.W. & Butterworth, R.F. 1997. Epidemic of seasonal ataxia in Nigeria following ingestion of the African silkworm *Anaphe venata*: Role of thiamine deficiency? *Metabolic Brain Disease*, 12(4): 251–258.

Ademolu, K.O., Idowu, A.B. & Olatunde, G.O. 2010. Nutritional value assessment of variegated grasshopper, *Zonocerus variegatus* (L.) (Acridoidea: Pygomorphidae), during post-embryonic development. *African Entomology*, 18(2): 360–364.

Adriaens, E.L. 1951. Recherches sur l'alimentation des populations au Kwango. *Bulletin Agricole du Congo Belge*, 42(2): 227–270.

Agea, J.G., Biryomumaisho, D., Buyinza, M. & Nabanoga, G.N. 2008. Commercialization of *Ruspolia nitidula* (Nsenene grasshoppers) in Central Uganda. *African Journal of Food Agriculture and Development*, 8(3): 319–332.

Agriprotein. 2010. Agriprotein. (available at www.agriprotein.com). Accessed October, 2012.

Aguilar-Miranda, E.D., Lopez, M.G., Escamilla-Santana, C. & Barba de la Rosa, A.P. 2002. Characteristics of maize flour tortilla supplemented with ground *Tenebrio molitor* larvae. *Journal of Agricultural and Food Chemistry*, 50(1): 192–195.

Akpalu, W., Muchapondwa, E. & Zikhali, P. 2009. Can the restrictive harvest period policy conserve mopane worms in southern Africa? A bioeconomic modelling approach. *Environment and Development Economics*, 14(5): 587–600.

Aldrich, J. 1988. Chemical ecology of the Heteroptera. *Annual Reviews of Entomology*, 33: 211–238.

Allotey, J. & Mpuchane, S. 2003. Utilization of useful insects as food source. *African Journal of Food, Agriculture, Nutrition and Development*, 3(2): 1–6.

Amadi, E.N., Ogbalu, O.K., Barimalaa, I.S. & Pius, M. 2005. Microbiology and nutritional composition of an edible larva (*Bunaea alcinoe* Stoll) of the Niger Delta. *Journal of Food Safety*, 25: 193–197.

Amar, Z. 2003. The Eating of locusts in Jewish tradition after the Talmudic period. *The Torah u-Madda Journal*, 11: 186–202.

Anand, H., Ganguly, A. & Haldar, P. 2008. Potential value of acridids as high protein supplement for poultry feed. *International Journal of Poultry Science*, 7(7): 722–725.

Anderson, S. J. 2000. Increasing calcium levels in cultured insects. *Zoo Biology*, 19(1): 1-9.

Auerswald, L. & Lopata, A. 2005. Insects: diversity and allergy. *Current Allergy & Clinical Immunology*, 18: 58–60.

Austin, A.D., Yeates, D.K., Cassis, G., Fletcher, M.J., La Salle, J., Lawrence, J.F., McQuillan, P.B., Mound, L.A., Bickel, D.J., Gullan, P.J., Hales, D.F. & Taylor, G.S. 2004. Insects "Down Under": diversity, endemism and evolution of the Australian insect fauna: examples from select orders. *Australian Journal of Entomology*, 43(3): 216–234.

Awoniyi, T.A.M., Adetuyi, F.C. & Akinyosoye, F.A. 2004. Microbiological investigation of maggot meal, stored for use as livestock feed component. *Journal of Food, Agriculture & Environment*, 2(3&4): 104–106.

Ayieko, M.A., Kinyuru, J.N., Ndong'a, M.F. & Kenji, G.M. 2012. Nutritional value and consumption of black ants (*Carebara vidua* Smith) from the Lake Victoria region in Kenya. *Advance Journal of Food Science and Technology*, 4(1): 39–45,

Ayieko, M.A., Ndong'a, M.F.O. & Tamale, A. 2010. Climate change and the abundance of edible insects in the Lake Victoria Region. *Journal of Cell and Animal Biology*, 4(7): 112–118.

Ayieko, M.A., Obonyo, G.O., Odhiambo, J.A., Ogweno, P.L., Achacha, J. & Anyango, J. 2011. Constructing and using a light trap harvester: rural technology for mass collection of agoro termites (*Macrotermes subhylanus*). *Research Journal of Applied Sciences, Engineering and Technology*, 3(2), 105–109.

Ayieko, M.A., Oriamo, V. & Nyambuga, I.A. 2010. Processed products of termites and lake flies: improving entomophagy for food security within the Lake Victoria region. *African Journal of Food, Agriculture, Nutrition and Development*, 10(2): 2085–2098.

Ayieko, M.A. & Oriaro, V. 2008. Consumption, indigeneous knowledge and cultural values of the lakefly species within the Lake Victoria region. *African Journal of Environmental Science and Technology*, 2(10): 282–286.

Bachstez, M. & Aragon, A. 1945. Notes on Mexican drugs, plants, and foods. III. Ahuauhtli, the Mexican caviar. *Journal of the American Pharmaceutical Association*, 34: 170–172.

BACSA. 2011. *Sericulture for multi products: new prospects for development*. Proceedings of the 5th BACSA International Conference, 11–15 April 2011, Bucharest, Romania. (available at www.bacsa-silk.org/user_pic/file/PROCEEDINGS%20SERIPRODEV%202011.pdf)

Bahuchet, S. 1975. Ethnozoologie des Pygmées Babinga de la Lobaye, République Centrafricaine. *In* R. Pujol, ed. *Premier Colloque d'Ethnozologie*. pp. 53–61. Paris, Institut International d'Ethnoscience.

Bahuchet, S. and Garine, L.D. 1990. Recipes for a forest menu. *In* C.M. Hladik, S. Bahuchet & L.D. Garine, eds. *Food and Nutrition in the African Rain Forest*. pp. 53–54. Paris, United Nations Educational, Scientific and Cultural Organization.

Baines, N. 2012. "Waiter, there's a fly in my soup": Insects as food are more than just a gastronomic gimmick. *The Independent*, online edition, posted on 7 September 2012. (available at www.independent.co.uk/life-style/food-and-drink/features/waiter-theres-a-fly-in-my-soup-insects-as-food-are-more-than-just-a-gastronomic-gimmick-8113939.html).

Balasubramanian, A. 1984. *Environmental economics – meaning, definition and importance: towards a philosophy of environmental education*. Singapore, Regional Institute of Higher Education and Development.

Banjo, A.D., Lawal, O.A. & Adeyemi, A.I. 2006. The microbial fauna associated with the larvae of *Oryctes monocerus*. *Journal of Applied Sciences Research*, 2(11): 837–843.

Banjo, A.D., Lawal, O.A. & Songonuga, E.A. 2006. The nutritional value of fourteen species of edible insects in southwestern Nigeria. *African Journal of Biotechnology*, 5(3): 298–301.

Barker, G. 2009. *The agricultural revolution in prehistory: why did foragers become farmers?* New York, USA, Oxford University Press.

Barletta, B. & Pini, C. 2003. Does occupational exposure to insects lead to species-specific sensitization? *Allergy*, 58: 868–870.

Barreteau, D. 1999. Les Mofu-Gudur et leurs criquets. *In* C. Baroin & J. Boutrais, eds. *L'homme et l'animal dans le bassin du lac Tchad: actes du colloque du reseau Mega-Tchad*. pp. 133–169. Paris, Institut de Recherche pour le Developpement, Université Nanterre.

BBC. 2004. Locusts rebranded as 'sky prawns'. (available at http://news.bbc.co.uk/2/hi/asia-pacific/4032143.stm). Accessed April 2012.

Behmer, S.T. 2006. Insect dietary needs: plants as food for insects. *Encyclopedia of Plant and Crop Science.* UK, Taylor & Francis.

Bennett, F.J. 1965. An inventory of the Kiganda foods. *The Uganda Journal*, 29(1): 45–53.

Bequaert, J. 1921. Insects as food. How they have augmented the food supply of mankind in early and recent years. *Natural History Journal,* 21: 191–200.

Bergier, E. 1941. *Peuples entomophages et insectes comestibles: étude sur les moeurs de l'homme et de l'insecte.* Avignon, Imprimerie Rulliere Freres.

Bodenheimer, F.S. 1951. *Insects as human food; a chapter of the ecology of man.* The Hague, Dr. W. Junk Publishers.

Bornemissza, G.F. 1976. The Australian Dung Beetle Project 1965–1975. *Australian Meat Research Committee Review*, 30: 1–30.

Boulidam, S. 2010. Edible insects in Lao market economy. *In* P.B. Durst, D.V. Johnson, R.L. Leslie. & K. Shono. *Forest insects as food: humans bite back, proceedings of a workshop on Asia-Pacific resources and their potential for development*, pp. 131–140. Bangkok, Thailand, FAO Regional Office for Asia and the Pacific.

Bouvier, G. 1945. Quelques questions d'entomologie vétérinaire et lutte contre certains arthropodes en Afrique tropicale. *Acta Trop*, 2: 42–59.

Bourne, G.H. 1953. The Food of the Australian Aboriginal. *Proceedings of the Nutrition Society*, 12: 58–65.

Bradbear, N. 2009. *Bees and their role in forest livelihoods: a guide to the services provided by bees and the sustainable harvesting, processing and marketing of their products.* Non-Wood Forest Products Series 19. Rome, FAO.

Brambell, F.W. 1965. Report of the Technical Committee to Enquire into the Welfare of Animals kept under Intensive Livestock Husbandry Systems. London, Her Majesty's Stationary Office, Great Britain. (also available at www.nhk.nl/downloads/1965_rapport_commissie_brambell.pdf)

Brinchmann, B.C., Bayat, M., Brøgger, T., Muttuvelu, D.V., Tjønneland, A. & Sigsgaard, T. 2011. A possible role of chitin in the pathogenesis of asthma and allergy. *Ann. Agric. Environ. Med.*, 18: 7–12.

Bukkens, S.G.F. 1997. The nutritional value of edible insects. *Ecology of Food and Nutrition*, 36: 287–319.

Bukkens, S.G.F. 2005. Insects in the human diet: nutritional aspects. *In* M.G. Paoletti, ed. *Ecological implications of minilivestock; role of rodents, frogs, snails, and insects for sustainable development*, pp. 545–577. New Hampshire, Science Publishers.

Burrows, C. 2012. Update regarding cochineal extract. In *Starbucks Blog*, posted on 29 March 2012 (available at http://blogs.starbucks.com/blogs/customer/archive/2012/03/29/update-regarding-cochineal-extract.aspx).

Campbell, M. 2011. Bug Appétit: San Francisco's Pre-Hispanic Snackeria. *The World*, posted on 2 November 2011 (available at www.theworld.org/2011/11/edible-bugs-food/).

Cerda, H., Martinez, R., Briceno, N., Pizzoferrato, L., Manzi, P., Tommaseo Ponzetta, M., Marin, O. & Paoletti, M.G. 2001. Palm worm (*Rhynchophorus palmarum*): traditional food in Amazonas, Venezuela. Nutritional composition, small scale production and tourist palatability. *Ecology of Food and Nutrition*, 40(1): 13–32.

Cerritos, R. 2009. Insects as food: an ecological, social and economical approach. *CAB Reviews: Perspectives in Agriculture, Veterinary Science, Nutrition and Natural Resources*, 4(27): 1–10.

Cerritos, R. & Cano-Santana, Z. 2008. Harvesting grasshoppers *Sphenarium purpurascens* in Mexico for human consumption: A comparison with insecticidal control for managing pest outbreaks. *Crop Protection*, 27(3-5): 473–480.

Césard, N. 2004a. Le kroto (*Oecophylla smaragdina*) dans la région de Malingping, Java-Ouest, Indonesie: collecte et commercialisation d'une resource animale non négligeable. *Anthropozoologica*, 39(2): 15–31.

Césard, N. 2004b. Harvesting and commercialisation of kroto (*Oecophylla smaragdina*) in the Malingping area, West Java, Indonesia. *In* K. Kusters & B. Belcher, eds. *Forest products, livelihoods and conservation: case studies of non-timber forest product systems. Volume 1, Asia*. pp. 61–78. Jakarta, Indonesia, CIFOR.

Chadwick, I. 2011. Mescal's history. (available at www.ianchadwick.com/tequila/mezcal_history.htm). Accessed Feburary 2013.

Chagnon, N.A. 1983. *Yanomamo: The fierce people*. New York, CBS College Publishing.

Chapagain, A.K. & Hoekstra, A.Y. 2003. *Virtual water flows between nations in relation to trade in livestock and livestock products*. Value of Water Research Report Series No. 13. Paris, United Nations Educational, Scientific and Cultural Organization.

Chavanduka, D.M. 1976. Insects as a source of protein to the Africain. *The Rhodesia Science News*, 9(7): 217–220.

Chen, P.P., Wongsiri, S., Jamyanya, T., Rinderer, T.E., Vongsamanode, S., Matsuka, M., Sylvester, H.A. & Oldroyd, B.P. 1998. Honey bees and other edible insects used as human food in Thailand. *American Entomologist*, 44(1): 24–28.

Chen, X., Feng, Y. & Chen, Z. 2009. Common edible insects and their utilization in China. *Entomological Research*, 39(5): 299–303.

Chen, Y.I. & Akre, R.D. 1994. Ants used as food and medicine in China. *The Food Insects Newsletter*, No. 7(2): 8.

Cherry, R. 1991. Use of insects by Australian aborigines. *American Entomologist*, 32: 8–13.

Chidumayo, E.N. & Mbata, K.J. 2002. Shifting cultivation, edible caterpillars and livelihoods in the Kopa area of northern Zambia. *Forests, Trees and Livelihoods*, 12: 175–193.

Choo, J. 2008. Potential ecological implications of human entomophagy by subsistence groups of the Neotropics. *Terrestrial Arthropod Reviews*, 1: 81–93.

Choo, J., Zent, E.L. & Simpson, B.B. 2009. The importance of traditional ecological knowledge for palm-weevil cultivation in the Venezuelan Amazon. *Journal of Ethnobiology*, 29(1): 113–128.

Chung, A.Y.C. 2010. Edible insects and entomophagy in Borneo. Edible insects in Lao market economy. *In* P.B. Durst, D.V. Johnson, R.L. Leslie. & K. Shono. *Forest insects as food: humans bite back, proceedings of a workshop on Asia-Pacific resources and their potential for development*, pp. 131–140. Bangkok, Thailand, FAO Regional Office for Asia and the Pacific.

Cohen, A.C. 2001. Formalizing Insect rearing and artificial diet technology. *American Entomologist*, 47(4): 199.

Cohen, J.H., Sánchez, N.D.M. & Montiel-ishinoet, F.D. 2009. Chapulines and food choices in rural Oaxaca. *Gastronomica: the Journal of Food and Culture*, 9(1): 61–65.

Coletto-Silva, A. 2005. Captura de enxames de abelhas sem ferrão (Hymenoptera, Apidae, Meliponinae) sem destruição de árvores. *Acta Amazonica*, 35(3): 383-388.

Collavo, A., Glew, R.H., Huang, Y.S., Chuang, L.T., Bosse, R. & Paoletti, M.G. 2005. House cricket small-scale farming. *In* M.G. Paoletti, ed., *Ecological implications of minilivestock: potential of insects, rodents, frogs and snails*. pp. 519–544. New Hampshire, Science Publishers.

Costa-Neto, E.M. 2003. Entertainment with insects: singing and fighting insects around the world. a brief review. *Etnobiología*, 3: 21–29.

Costa-Neto, E.M. 2012. Estudos etnoentomológicos no estado da Bahia, Brasil: uma homenagem aos 50 anos do campo de pesquisa. *Biotemas*, 17(1): 117–149.

Costermans, J.B. 1955. Het termieten-stoken bij de Logo-Avokaya (vervolg). *Aequatoria*, 18(2): 50–55.

Crook R.J. & Walters, E.T. 2011. Nociceptive behavior and physiology of mollusks: animal welfare implications. *ILAR J*, 52: 185–195.

Cutter, C.N. 2006. Opportunities for bio-based packaging technologies to improve the quality and safety of fresh and further processed muscle foods. *Meat Science*, 74(1): 131–142.

Das, M., Ganguly, A. & Haldar, P. 2009. Space requirement for mass rearing of two common Indian acridid adults (Orthoptera: Acrididae) in laboratory condition. *American-Eurasian J. Agric. & Environ. Sci.* 6(3): 313–316.

Das, M., Ganguly, A. & Haldar, P. 2010. Nutrient analysis of grasshopper manure for soil fertility enhancement. *American-Eurasian J. Agric. & Environ. Sci.* 7(6): 671–675.

Davey, G.C.L. 1994. The "disgusting" spider: The role of disease and illness in the perpetuation of fear of spiders. *Society and Animals*, (2): 1.

Davis, G.R.F. & Sosulski, F.W. 1974. Nutritional quality of oilseed protein isolates as determined with larvae of the yellow mealworm, *Tenebrio molitor* L. *The Journal of Nutrition*, 104(9): 1172–1177.

Decary, R. 1937. L'entomophagie chez les indigènes de Madagascar. *Bulletin de la Société entomologique de France* (9 Juin 1937), pp. 168–171.

DeFoliart, G.R. 1989. The human use of insects as food and as animal feed. *Bulletin of the Entomological Society of America*, 35: 22–35.

DeFoliart, G.R. 1995. Edible insects as minilivestock. *Biodiversity and Conservation*, 4(3): 306–321.

DeFoliart, G.R. 1997. An overview of the role of edible insects in preserving biodiversity. *Ecology of Food and Nutrition*, 36(2–4): 109–132.

DeFoliart, G.R. 1999. Insects as food: Why the western attitude is important. *Annual Review of Entomology*, 44: 21–50.

DeFoliart, G.R. 2002. *The human use of insects as food resource: a bibliographic account in progress.* Wisconsin, USA, Department of Entomology, University of Wisconsin-Madison. (also availabe at: www.food-insects.com/book7_31/The%20Human%20 Use%20of%20Insects%20as%20a%20Food%20Resource.htm)

DeFoliart, G.R. 2005. An overview of role of edible insects in preserving biodiversity. *In* M.G. Paoletti, ed., *Ecological implications of minilivestock: potential of insects, rodents, frogs and snails.* pp. 123–140. New Hampshire, USA, Science Publishers.

DeFoliart, G., Dunkel, F.V. & Gracer, D. 2009. *The Food Insects Newsletter.* Salt Lake City, Utah, USA. Aardvark Global Publishing.

Del Toro, I., Ribbons, R.R. & Pelini, S.L. 2012. The little things that run the world revisited: a review of ant-mediated ecosystem services and disservices (Hymenoptera: Formicidae). *Myrmecological News*, 17: 133–146.

Delong, D.M. 1960. Man in a world of insects. *The Ohio Journal of Science*, 60(4): 193–206.

Diamond, J. 2005. *Guns, germs and steel: a short history of everybody for the last 13 000 years.* UK, Vintage.

Dufour, D.L. 1987. Insects as food: a case study from the northwest Amazon. *American Anthropologist*, 89(2): 383.

Durst, P.B. & Shono, K. 2010. Edible forest insects: exploring new horizons and traditional practices. *In* P.B. Durst, D.V. Johnson, R.L. Leslie. & K. Shono. *Forest insects as food: humans bite back, proceedings of a workshop on Asia-Pacific resources and their potential for development*, pp. 1–4. Bangkok, FAO Regional Office for Asia and the Pacific.

Dyson-Hudson, R. & Smith, E.A. 1978. Human territoriality: an ecological reassessment. *American Anthropologist*, 80(1): 21–41.

Dzerefos, C.M., Witkowski, E.T.F. & Toms, R. 2009. Life-history traits of the edible stinkbug, *Encosternum delegorguei* (Hem., Tessaratomidae), a traditional food in southern Africa. *J. Appl. Entomol.*, 133: 749–759.

Egert, M., Wagner, B., Lemke, T., Brune, A. & Friedrich, M.W. 2003. Microbial community structure in midgut and hindgut of the humus-feeding larva of *Pachnoda ephippiata* (Coleoptera: Scarabaeidae). *Applied and Environmental Microbiology*, 69(11): 6659–6668.

Eisemann, C.H., Jorgensen W.K., Merritt, D.J, Rice, M.J., Cribb, B.W, Webb, P.D. & Zalucki, M.P. 1984. Do insects feel pain? A biological view. *Experientia*, 40: 164–167.

Ekoue, S.K. & Hadzi, Y.A. 2000. Production d'asticots comme source de protéines pour jeunes volailles au Togo: observations préliminaires. *Tropicultura*, 18(4): 212–214.

El-Mallakh, O.S. & El-Mallakh, R.S. 1994. Insects of the Qur'an (Koran). *American Entomologist*, 40: 82–84.

Elvin, C.M., Carr, A.G., Huson, M.G., Maxwell, J.M., Pearson, R.D., Vuocolo, T., Liyou, N.E., Wong, D.C.C., Meritt, D.J. & Dixon, N.E. 2005. Synthesis and properties of crosslinked recombinant pro-resilin. *Nature*, 437: 999–1002.

Ento. 2012. Available at http://cargocollective.com/ento/. Accessed September 2012.

Erens, J., Es van, S., Haverkort, F., Kapsomenou, E. & Luijben, A. 2012. *A bug's life: large-scale insect rearing in relation to animal welfare: Project 1052*. Wageningen, Wageningen University.

Erickson, M.C., Islam, M., Sheppard, C., Liao, J. & Doyle, M.P. 2004. Reduction of *Escherichia coli* O157:H7 and *Salmonella enterica* serovar Enteritidis in chicken manure by larvae of the black soldier fly. *J. Food Prot.*, 67(4): 685–690.

Ernst, W.H.O. & Sekhwela, M.B.M. 1987. The chemical composition of lerps from the mopane psyllid *Arytaina mopane* (Homoptera, Psyllidae). *Insect Biochem.*, 17(6): 905–909.

European Commission. 1997. *Regulation (EC) No 258/97 of the European Parliament and of the Council of 27 January 1997 concerning novel foods and novel food ingredients*. Brussels.

European Comission. 2008. Ecodoptera Project (available at: http://ec.europa.eu/environment/life/project/Projects/index.cfm?fuseaction=home.createPage&s_ref=LIFE05%20ENV/E/000302&area=2&yr=2005&n_proj_id=2897&cfid=16586&cftoken=2e4adf8baa61f2ac-360A2F1D-DAE5-7FE0-A7720CC7129F3210&mode=print&menu=false).

Fairman, R.J. 2010. Instigating an education in insects: the eating creepy crawlies' exhibition. *Antenna*, 34: 169–170.

FAO. 1997a. *Guidelines for small-scale fruit and vegetable processors*. FAO Agricultural Services Bulletin 127. Rome.

FAO. 1997b. *Rural aquaculture: overview and framework for country reviews*. Rome.

FAO. 2003. *State of forest and tree genetic resources in dry zone Southern Africa Development Community countries*. Rome.

FAO. 2004. *Contribution des insectes de la forêt à la sécurité alimentaire: L'exemple des chenilles d'Afrique Centrale*. NTFP Working document No. 1. FAO, Rome. (also available at www.fao.org/docrep/007/j3463f/j3463f3400.htm).

FAO. 2007. *Promises and challenges of the informal food sector in developing countries*. Rome.

FAO. 2009a. How to feed the world in 2050. Paper presented at the High Level Expert Forum, Rome, Italy, 12–13 October. (available at www.fao.org/fileadmin/templates/wsfs/docs/expert_paper/How_to_Feed_the_World_in_2050.pdf).

FAO. 2009b. *Biodiversity and nutrition, a common path*. Rome.

FAO. 2010a. *Development of regional standard for edible crickets and their products*. Paper presented at the Joint FAO/WHO meeting Food Standards Programme: FAO/WHO Coordinating Committee for Asia, Bali, Indonesia.

FAO. 2010b. *Biodiversity and sustainable diets: united against hunger*. Report presented at World Food Day/World Feed Week, 2–5 November, 2010, Rome.

FAO. 2010c. *Expert consultation on nutrition indicators for biodiversity: Volume 2*. Rome.

FAO. 2011a. *Selling street and snack foods*. Rome.

FAO. 2011b. *Small farm for small animals*. Rome.

FAO. 2011c. *State of food and agriculture 2010-2011. Women in agriculture: closing the gender gap for development*. Rome.

FAO. 2012a. *State of the world's forests 2012*. Rome

FAO. 2012b. *State of the world fisheries*. Rome.

FAO. 2012c. Water & poverty, an issue of life & livelihoods. (available at www.fao.org/nr/water/issues/scarcity.html). Accessed November, 2012.

FAO. 2012d. FAO at Rio +20. (available at www.fao.org/rioplus20/en/). Accessed January 2013.

FAO. 2012e. Gender and nutrition. (available at www.fao.org/docrep/012/al184e/al184e00.pdf). Accessed on November, 2012.

FAO. 2012f. Composition database for Biodiversity Version 2, BioFoodComp2. (Latest update: 10 January 2013). Accessed January 2012. (available at www.fao.org/infoods/infoods/tables-and-databases/en/).

FAO/IAEA. 2001. *Manual on the application of the HACCP system in Mycotoxin prevention and control*. FAO Food and Nutrition paper, No. 73. Rome, FAO/IAEA Training and Reference Centre for Food and Pesticide Control.

FAO/WHO. 2001a. *Codex Alimentarius: Joint FAO/WHO Food Standards Programme*. Rome. (available at www.codexalimentarius.org/).

FAO/WHO. 2001b. *Human vitamin and mineral requirements*. Rome.

FAO/WUR. 2012. *Expert consultation meeting: assessing the potential of insects as food and feed in assuring food security*. P. Vantomme, E. Mertens, A. van Huis & H. Klunder, eds. Summary report, 23–25 January 2012, Rome. Rome, FAO.

Farina, L. , Demey, F. & Hardouin, J. 1991. Production de termites pour l'aviculture villageoise au Togo. *Tropicultura*, 9(4): 181–187.

Fasoranti, J.O. & Ajiboye, D.O. 1993. Some edible insects of Kwara State, Nigeria. *American Entomologist*, 39(2): 113–116.

Faure, J.C. 1944. Pentatomid bugs as human food. *Journal Ent. Soc. S. Africa*, 7: 110–112.

FDA. 2011. **Defect levels handbook.** (available at www.fda.gov/food/guidancecompliance regulatoryinformation/guidancedocuments/sanitation/ucm056174.htm). Accessed January 2013.

Feng, Y. & Chen, X. 2003. Utilization and perspective of edible insects in China. *Forest Science and Technology*, 44(4): 19–20.

Feng, Y. & Chen, X. 2009. Reviews on the research and utilization of insect health care foods. *Journal of Shandong Agrcultural University (Natural science)*, 40(4): 676–680.

Feng, Y., Zhao, M., He, Z., Chen, Z. & Sun, L. 2009. Research and utilization of medicinal insects in China. *Entomological Research*, 39(5): 313–316.

Fessler, D.M.T. & Navarette, C.D. 2003. Meat is good to taboo: Dietary proscriptions as a product of the interaction of psychological mechanisms and social processes. *Journal of Cognition and Culture*, 3(1): 1–40.

Fiala, N. 2008. Meeting the demand: an estimation of potential future greenhouse gas emissions from meat production. *Ecological Economics*, 67: 412–419.

Finke, M.D. 2002. Complete nutrient composition of commercially raised invertebrates used as food for insectivores. *Zoo Biology*, 21(3): 269–285.

Finke, M.D. 2005. Nutrient composition of bee brood and its potential as human food. *Ecology of Food and Nutrition*, 44(4), 257–270.

Finke, M.D. 2007. Estimate of chitin in raw whole insects. *Zoo Biology*, 26, 105–115.

Flood, J. 1980. *The moth hunters: Aboriginal prehistory of the Australian Alps*. Canberra, Humanities Press, Inc.

Friedland, S.R. 2007. *Food and morality: proceedings of the Oxford Symposium on Food and Cookery*. Prospect Books.

Gaston, K.J. & Chown, S.L. 1999. Elevation and climatic tolerance: a test using dung beetles. *Oikos*, 86: 584–590.

Ghazoul, J. 2006. *Mopani woodlands and the mopane worm: enhancing rural livelihoods and resource sustainability. Final technical report*. London, DFID.

Ghesquière, J. 1947. Les insectes palmicoles comestibles. P. Lepesme, ed. *Les insectes des palmiers*, pp. 791–793. Paris, P. Lechevalier.

Giaccone, V. 2005. Hygiene and health features of "minilivestock". *In* M.G. Paoletti, ed. *Ecological implications of minilivestock; role of rodents, frogs, snails, and insects for sustainable development*, pp. 579–598. New Hampshire, Science Publishers.

Giblin-Davis, R.M., Gerber, K., & Griffith, R. 1989. Laboratory rearing of the red palm weevil, *Rhynchophorus cruentatus* and *R. palmarum* (Coleoptera: Curculionidae). *Florida Entomologist*, 72(3): 480–488.

Gitta, A. 2012. Power cuts harm Uganda's grasshopper business. *Deutsche Welle*, online edition. Posted on 3 January 2012 (available at www.dw.de/power-cuts-harm-ugandas-grasshopper-business/a-15634688-1).

Glew, R.H., Jackson, D., Sena, L., VanderJagt, D.J., Pastuszyn, A., & Millson, M. 1999. *Gonimbrasia belina* (Lepidoptera: Saturniidae), a nutritional food source rich in protein, fatty acids and minerals. *American Entomologist* (winter 1999), 45(4): 250–253.

Godfray, H.C.J. 1994. *Parasitoids: behavioural and evolutionary ecology*. Princeton, USA, Princeton University Press.

Gomez, P.A., Halut, R. & Collin, A. 1961. Production de proteines animales au Congo. *Bulletin Agricole du Congo*, 52(4): 689–786.

Gon, S.M., & Price, E.O. 1984. Invertebrate domestication: behavioral considerations. *BioScience*, 34(9): 575–579.

Goodman, W.G. 1989. Chitin: A magic bullet? *The Food Insects Newsletter*, 2(3): 1, 6–7.

Gottschling, S. & Meyer, S. 2006. An edpidemic airborne disease cause by the oak processionary caterpillar. *Pediatric Dermatology*, 23(1): 64–66.

Gounari, S. 2006. Studies on the phenology of *Marchalina hellenica* (gen.) (Hemiptera: Coccoidea, Margarodidae) in relation to honeydew flow. *Journal of apicultural research*, 45(1): 8–12.

Government of India. 2011. *Central Silk Board annual report 2010–2011*. Ministry of Textiles.

Green, K., Broome, L., Heinze, D. & Johnston, S. 2001. Long distance transport of arsenic by migrating bogong moths from agricultural lowlands to mountain ecosystems. *The Victorian Naturalist*, 118(4): 112–116.

Groot, A. 1995. *La protection des végétaux dans les cultures de subsistance: le cas du mil au Niger de l' Ouest*. Wageningen, the Netherlands, Wageningen University.

Guénard, B., Weiser, M. & McCoy, N. 2010. *Ant genera of the world: Oecophylla*. (available at www.antmacroecology.org). Accessed April, 2012.

Guerin-Meneville, M.F.E. 1857. Entomologie appliquée. Mémoire sur trois espèces d'insectes hémiptères du groupe des punaises aquatiques, dont les œufs servent à faire une sorte de pain nomme hautle, au Mexique. *Bulletin Societé Imperiale Zoologique d'Acclimatation*, 4: 578–581.

Hackstein, J.H. & Stumm, C.K. 1994. Methane production in terrestrial arthropods. *Proceedings of the National Academy of Sciences of the United States of America*, 91(12): 5441–5445.

Hale, O.M. 1973. Dried *Hermetia illucens* larvae (Stratiomyidae) as a feed additive for poultry. *Journal of the Georgia Entomological Society*, 8: 16–20.

Haldar, P. 2012. *Evaluation of nutritional value of short-horn grasshoppers (acridids) and their farm-based mass production as a possible alternative protein source for human and livestock*. Paper presented at the Expert Consultation Meeting on Assessing the Potential of Insects as Food and Feed in assuring Food Security, 23–26 January, Rome, FAO.

Hanboonsong, Y. 2010. Edible insects and associated food habits in Thailand. *In* P.B. Durst, D.V. Johnson, R.L. Leslie. & K. Shono. *Forest insects as food: humans bite back, proceedings of a workshop on Asia-Pacific resources and their potential for development*, pp. 171–182. Bangkok, FAO Regional Office for Asia and the Pacific.

Hanboonsong, Y. 2012. *Edible insect recipes: edible insects for better nutrition and improved food security*. Vientiane, FAO and the Government of Lao PDR.

Handley, M.A. 2007. Globalization, binational communities, and imported food risks: results of an outbreak investigation of lead poisoning in Monterey County, California. *American Journal of Public Health*, 97(5): 900–906.

HaoCheng Mealworm, Inc. 2012. About HaoCheng Mealworm Inc. (available at www.hcmealworm.com). Accessed November 2012.

Hardouin, J. 1995. Minilivestock: from gathering to controlled production. *Biodiversity and Conservation*, 4(3): 220–232.

Harpe, D. & McCormack, D. 2001. Online Etymological Dictionary. (available at www.LogoBee.com). Accessed on 1 November 2012.

Harris, W. V. 1940. Some notes on insects as food. *Tanganyika Notes and Records*, 9: 45–48.

Harrison, S.G. 1950. Manna and its sources. *Kew Bulletin*, 5(3): 407–417.

Headings, M.E. & Rahnema, S. 2002. The nutritional value of mopane worms, *Gonimbrasia belina* (Lepidoptera: Saturniidae) for human consumption. Presentation at the Ten-Minute Papers: Section B. Physiology, Biochemistry, Toxicology and Molecular Biology Series, 20 November 2002. Ohio, USA, Ohio State University.

Health Canada. 2006. Food additives permitted for use in Canada. (available at www.hc-sc.gc.ca/fn-an/securit/addit/diction/dict_food-alim_add-eng.php). Accessed November, 2012.

Heinrichs, E.A. & Mochida, O. 1984. From secondary to major pest status: the case of insecticide-induced rice brown planthopper *Nilaparvata lugens*, resurgence. *Protection Ecology*, 7(2-3): 201–218.

Heinz, G. & Hautzinger, P. 2007. *Meat processing technology for small- to medium-scale producers*. FAO Regional Office for Asia and the Pacific. (also available at www.fao.org/docrep/010/ai407e/ai407e00.htm).

Herz, R. 2012. *That's disgusting: unraveling the mysteries of repulsion*. New York, USA, W.W. Norton & Co.

Hobane, P.A. 1994. The urban marketing of the mopane worm: the case of Harare. *CASS – NRM Occasional Paper Series*. Harare, Zimbabwe, Centre for Applied Social Sciences, University of Zimbabwe.

Hogue, C.L. 1987. Cultural entomology [Review]. *Annual Review of Entomology*, 32: 181–199.

Holden, S. 1991. Edible caterpillars: a potential agroforestry resource? They are appreciated by local people, neglected by scientists. *The Food Insects Newsletter*, 4(2), 3–4. USA.

Hölldobler, B. 1983. Territorial behaviour in the green tree ant *Oecophylla smaragdina*. *Biotropica*, 15(4): 241–250.

Hölldobler, B. & Wilson, E.O. 2010. *The leafcutter ants: civilization by instinct*. New York, USA, W. W. Norton & Company.

Holt, V.M. 1995. *Why not eat insects?* Oxford, UK, Thornton's.

Hope, R.A., Frost, P.G.H., Gardiner, A. & Ghazoul, J. 2009. Experimental analysis of adoption of domestic mopane worm farming technology in Zimbabwe. *Development Southern Africa*, 26: 29–46.

Hou, L., Shi, Y., Zhai, P., & Le, G. 2007. Inhibition of foodborne pathogens by Hf-1, a novel antibacterial peptide from the larvae of the housefly (*Musca domestica*) in medium and orange juice. *Food Control*, 18(11): 1350–1357.

Howard, R.W. & Stanley-Samuelson, D.W. 1990. Phospholipid fatty acid composition and arachidonic acid metabolism in selected tissues of adult *Tenebrio molitor* (Coleoptera: Tenebrionidae). *Annals of the Entomological Society of America*, 83(5): 975–981.

Hrabar, H., Hattas, D. & Toit, J.T. 2009. Differential effects of defoliation by mopane caterpillars and pruning by African elephants on the regrowth of *Colophospermum mopane* foliage. *Journal of Tropical Ecology*, 25: 301–309.

Hu, E.Z., Bartsev, S.I. & Liu, H. 2010. Conceptual design of a bioregenerative life support system containing crops and silkworms. *Advances in Space Research*, 45(7): 929–939.

Hwangbo, J., Hong, E.C., Jang, A., Kang, H.K., Oh, J.S., Kim, B.W. & Park, B.S. 2009. Utilization of house fly-maggots, a feed supplement in the production of broiler chickens. *J. Environ Biol.*, 30(4): 609–614.

IFIF. 2012. International Feed Industry Federation. (available at www.ifif.org). Accessed May 2012.

Ijaiya, A.T. & Eko, E.O. 2009. Effect of replacing dietary fish meal with silkworm (*Anaphe infracta*) caterpillar meal on performance, carcass characteristics and haematological parameters of finishing broiler chicken. *Pakistan Journal of Nutrition*, 8(6): 850–855.

Illgner, P. & Nel, E. 2000. The geography of edible insects in sub-Saharan Africa: a study of the mopane caterpillar. *Geographical Journal*, 166: 336–351.

Ingram, M., Nabhan, G.P. & Buchmann, S. L. 1996. Our forgotten pollinators: protecting the brids and bees. *Global Pesticide Campaigner*, 6(4): 1–12.

Institute of Food Technologists. 2011. Developing solutions for developing countries. (available at www.ift.org/community/students/competitions/developing-solutions-for-developing-countries.aspx). Accessed December 2012.

IPCC. 2007. Summary for policymakers. *In* S. Solomon, D. Qin, M. Manning, Z. Chen, M. Marquis, K.B. Averyt, M.Tignor & H.L. Miller, eds. *Climate change 2007: The physical science basis. Contribution of Working Group I to the fourth assessment report of the Intergovernmental Panel on Climate Change*. Cambridge, UK & New York, USA, Cambridge University Press.

Iroko, A.F. 1982. Le rôle des termitières dans l'histoire des peuples de la République Populaire du Bénin des origines a nos jours. *Bulletin de l'I.F.A.N*, 44: 50–75.

ISC. 2011. *22nd Congress of the International Sericultural Commision*. Bangalore, India. (available at www.qsds.go.th/isccongress/files/Proceeding.pdf)

Janzen, D.H. & Schoener, T.W. 1968. Differences in insect abundance and diversity between wetter and drier sites during tropical dry season. *Ecology*, 49: 96–110.

Jensen, R. L., Newsom, L.D., Herzog, D.C., Thomas, J.W., Farthing, B.R. & Martin, F.A. 1977. A method of estimating insect defoliation of soybean. *Journal of Economic Entomology*, 70(2): 240–242.

Jin, X.B. & Yen, A.L. 1998. Conservation and the cricket culture in China. *Journal of Insect Conservation*, 2: 211–216.

Johnson, D.V. 2010. The contribution of edible forest insects to human nutrition and to forest management. *In* P.B. Durst, D.V. Johnson, R.L. Leslie. & K. Shono. *Forest insects as food: humans bite back, proceedings of a workshop on resources and their potential for development*, pp. 5–22. Bangkok, FAO Regional Office for Asia and the Pacific.

Johnson, N. & Berdegué, J. 2004. Property rights, collective action and agribusiness. *2020 Focus*, 11: 13.

Jongema, Y. 2012. List of edible insect species of the world. Wageningen, Laboratory of Entomology, Wageningen University. (available at www.ent.wur.nl/UK/Edible+insects/Worldwide+species+list/).

Kampmeier, G.E. & Irwin, M.E. 2003. Commercialization of insects and their products. *In* V.H. Resh & R.T. Cardé, eds. *Encyclopedia of insects*, pp. 252–260. Burlington, USA, Academic Press.

Katayama, N., Ishikawa, Y., Takaoki, M., Yamashita, M., Nakayama, S., Kiguchi, K., Kok, R., Wada, H. & Mitsuhashi, J. 2008. Entomophagy: a key to space agriculture. *Advances in Space Research*, 41(5): 701–705.

Katayama, N., Yamashita, M., Wada, H. & Mitsuhashi, J. 2005. Entomophagy as part of a space diet for habitation on Mars. *J. Space Tech. Sci.*, 21(2): 27–38.

Kellert, S.R. 1993. Values and perceptions of invertebrates. *Conservation Biology*, 7(4): 845–855.

Kenis, M., Sileshi, G., Mbata, K., Chidumayo, E., Meke, G. & Muatinte, B. 2006. Towards conservation and sustainable utilization of edible caterpillars of the miombo. Presentation to the SIL Annual Conference on Trees for Poverty Alleviation, 9 June 2006, Zürich, Switzerland.

Kinyuru, J.N., Kenji, G.M. & Muhoho, S.N. 2010. Nutritional potential of longhorn grasshopper (*Ruspolia differens*) consumed in Siaya District, Kenya. *Journal of Agriculture, Science and Technology*, 12(1): 1–24.

Kinyuru, J.N., Kenji, G.M. & Njoroge, M.S. 2009. Process development, nutrition and sensory qualities of wheat buns enriched with edible termites (*Macrotermes subhyalinus*) from Lake Victoria region, Kenya. *African Journal of Food and Agriculture Nutrition and Development*, 9(8): 1739–1750.

Kirkpatrick, T.W. 1957. *Insect life in the tropics*. London, Longmans, Green.

Kitsa, K. 1989. Contribution des insectes comestibles a l'amélioration de la ration alimentaire au Kasai-Occidental. *Zaire-Afrique*, 239: 511–519.

Klunder, H.C., Wolkers-Rooijackers, J., Korpela, J.M. & Nout, M.J.R. 2012. Microbiological aspects of processing and storage of edible insects. *Food Control*, 26: 628–631.

Kogan, M. 1998. Integrated pest management: historical perspectives and contemporary developments. *Annual Review of Entomology*, 43(1): 243–270.

Kok, R. 1983. The production of insects for human food. *Can. Inst. Food Sci. Technol. J.*, 16(1): 5–18.

Kok, R., Shivhare, U.S. & Lomaliza, K.J. 1990. Mass and component balances for insect production. *Canadian Agricultural Engineering*, 33(1): 185–192.

Kozanayi, W. & Frost, P. 2002. *Marketing of mopane worm in Southern Zimbabwe*. Harare, Institute of Environmental Studies.

Krishnan, R., Sherin, L., Muthuswami, M., Balagopal, R. & Jayanthi, C. 2011. Seri waste as feed substitute for broiler production. *Sericologia*, 51(3): 369–377.

Kuhnlein, H.V., Erasmus, B. & Spigelski, D. 2009. *Indigenous peoples' food systems: the many dimensions of culture, diversity and environment for nutrition and health*. FAO & Centre for Indigenous Peoples' Nutrition and Environment, Rome.

Kuyten, P. 1960. Darmafsluiting veroorzaakt door het eten van kevers. *Entomologische berichten*, 20(8): 143.

Lähteenmäki-Uutela, A. 2007. European novel food legislation as a restriction to trade. Paper presented at the seminar Pro-poor Development in Low-income Countries, 25–27 October, Montpellier, France. (also available at http://ageconsearch.umn.edu/bitstream/7909/1/pp07la01.pdf).

Latham, P. 1999. Edible caterpillars of the Bas Congo region of the Democratic Republic of Congo. *Antenna*, 23(3): 135–139.

Latham, P. 2003. *Edible caterpillars and their food plants in Bas-Congo*. Canterbury, Mystole Publications.

Lee, K.P., Simpson, S. J. & Wilson, K. 2008. Dietary protein-quality influences melanization and immune function in an insect. *Functional Ecology*, 22(6): 1052–1061.

Leung, W. 2012. Crushed bugs give Starbucks Frappucino its pretty pink colour. *The Globe and Mail*, online edition, posted on 28 March 2012 (available at www.theglobeandmail.com/life/the-hot-button/crushed-bugs-give-starbucks-frappucino-its-pretty-pink-colour/article621424/

Lewis, V.L. 1992. Spider silk: the unraveling of a mystery. *Acc. Chem. Res.*, 25: 392–398.

Li, Q., Zheng, L., Cai, H., Garza, E., Yu, Z. & Zhou, S. 2011. From organic waste to biodiesel: black soldier fly, *Hermetia illucens*, makes it feasible. *Fuel*, 90: 1545–1548.

Lindqvist, L. & Block, M. 1995. Excretion of cadmium during moulting and metamorphosis in *Tenebrio molitor* (Coleoptera; Tenebrionidae). *Comparative Biochemistry and Physiology*, 111(2): 325–328.

Liu, Q., Tomberlin, J.K., Brady, J.A., Sanford, M.R. & Yu, Z. 2008. Black soldier fly (Diptera: Stratiomyidae) larvae reduce *Escherichia coli* in dairy manure. *Environ. Entomol.*, 37(6): 1525–1530.

Lockwood, J.A. 2004. *Locust: the devastating rise and disappearance of the insect that shaped the American frontier.* New York, USA, Basic Books.

Lokkers, C. 1990. *Colony dynamics of the green tree ant (*Oecophylla smaragdina *Fab.) in a seasonal tropical climate.* Rockhampton, Australia, James Cook University of North Queensland.

Looy, H. & Wood, J.R. 2006. Attitudes toward invertebrates: are educational "bug banquets" effective? *The Journal of Environmental Education*, 37(2): 37–48.

Losey, J.E. & Vaughan, M. 2006. The economic value of ecological services provided by insects. *BioScience*, 56(4): 311–323.

Madsen, D.B. & Kirkman, J.E. 1988. Hunting hoppers. *American Antiquity*, 53(3): 593–604.

Makuku, S.J. 1993. All this for a bug! Community approaches to common property resources management: the case of the Norumedzo community in Bikita, Zimbabwe. *Forests, Trees and People Newsletter*, 22.

Malaisse. 1997. *Se nourir en foret claire africaine: approche écologique et nutritionnelle.* Gembloux, Les Presses Agronomiques de Gembloux.

Mariod, A., Klupsch, S., Hussein, I.H. & Ondruschka, B. 2006. Synthesis of alkyl esters from three unconventional sudanese oils for their use as biodiesel. *Energy & Fuels*, 20: 2249–2252.

Mariod, A., Matthäus, B. & Eichner, K. 2004. Fatty acid, tocopherol and sterol composition as well as oxidative stability of three unusual sudanese oils. *Journal of Food Lipids*, 11: 179–189.

Mariod, A., Matthäus, B., Eichner, K. & Hussein, I.H. 2005. Improving the oxidative stability of sunflower oil by blending with *Sclerocarya birrea* and *Aspongopus viduatus* oils. *Journal of Food Lipids*, 12: 150–158.

Mbata, K.J. & Chidumayo, E.N. 2003. Traditional values of caterpillars (Insecta: Lepidoptera) among the Bisa people of Zambia. *Insect Sci. Applic.*, 23(4): 341–354.

Mbata, K.J., Chidumayo, E.N. & Lwatula, C.M. 2002. Traditional regulation of edible caterpillar exploitation in the Kopa area of Mpika district in northern Zambia. *Journal of Insect Conservation*, 6(2): 115–130.

McCrae, A.W.R. 1982. Characteristics of swarming in the African edible bush-cricket *Ruspolia differens* (Serville) (Orthoptera, Tettigonioidea). *Journal of the East Africa Natural History Society and National Museum*, 178: 1–5.

Mela, D.J. 1999. Food choice and intake: the human factor. *Proceedings of the Nutrition Society*, 58: 513–521.

Menzel, C. 2002. *The lychee crop in the Asia and the Pacific.* FAO/RAP Publication 16. Bangkok, FAO Regional Office for Asia and the Pacific.

Mercer, C.W.L. 1994. Sago grub production in Lubu swamp near Lae-Papua New Guinea. *Klinkii*, 5(2): 30–34.

Mercer, C.W.L. 1997. Sustainable production of insects foe food and income by New Guinea villagers. *Ecology of Food and Nutrition*, 36(2–4): 151–157.

Meslin, F.X. & Formenty, P. 2004. A review of emerging zoonoses and the public health implications. Report of the WHO/FAO/OIE joint consultation on emerging zoonotic diseases, 3–5 May 2004, Geneva.

Meuwissen, P. 2011. *Insecten als nieuwe eiwitbron: Een scenarioverkenning van de marktkansen.* 's-Hertogenbosch: ZLTO projecten.

Meyer-Rochow, V.B. 1979. The diverse uses of insects in traditional societies. *Ethnomedicine*, 5(3/4): 287–300.

Meyer-Rochow, V.B. 2005. Traditional food insects and spiders in several ethnic groups of northeast India, Papua New Guinea, Australia and New Zealand. *In* M.G. Paoletti,

ed. *Ecological implications of minilivestock; role of rodents, frogs, snails, and insects for sustainable development*, pp. 385–409. New Hampshire, USA, Science Publishers.
Michaelsen, K.F., Hoppe, C., Roos, N., Kaestel, P., Stougaard, M., Lauritzen, L. & Mølgaard, C. 2009. Choice of foods and ingredients for moderately malnourished children 6 months to 5 years of age. *Food and Nutrition Bulletin*, 30(3): 343–404.
Mignon, J. 2002. L'entomophagie: une question de culture? *Tropicultura*, 20(3): 151–155.
Milton, K. 1984. Protein and carbohydrate resources of the Maku Indians of northwestern Amazonia. *American Anthropologist*, 86(1): 7–27.
Mitsuhashi, J. 2005. Edible insects in Japan. *In* M.G. Paoletti, ed. *Ecological implications of minilivestock; role of rodents, frogs, snails, and insects for sustainable development*, pp. 251–262. New Hampshire, USA, Science Publishers.
Mormino, V. 2009. Evil weevils attack Sicily! *Best of Sicily*. (available at www.bestofsicily.com/mag/art325.htm). Accessed November 2012.
Morris, B. 2004. *Insects and human life*. Oxford, UK, Berg.
Mors, P.O. 1958. Grasshoppers as food in Buhaya. *Anthropological Quarterly*, 31(2): 56–58.
Motte-Florac, É. & Thomas, J.M.C. 2003. Les "insectes" dans la tradition orale. *Ethnosciences*, 11: 407.
Mpuchane, S., Taligoola, H.K. & Gashe, B.A. 1996. Fungi associates with *Imbrasia belina*, an edible grasshopper. *Botswana Notes and Records*, 28: 193–197.
Munthali, S.M. & Mughogho, D.E.C. 1992. Economic incentives for conservation: bee-keeping and Saturniidae caterpillar utilization by rural communities. *Biodiversity and Conservation*, 1: 153–154.
Munyuli Bin Mushambanyi, T. 2000. Étude préliminaire orientée vers la production des chenilles comsommables par l'élevage des papillons (*Anaphe infracta*: Thaumetopoeidae) à Lwiro, Sud-Kivu. République Démocratique du Congo. *Tropicultura*, 18(4): 208–211.
Munyuli Bin Mushambanyi, T. & Balezi, N. 2002. Utilisation des blattes et des termites comme substituts potentiels de la farine de viande dans l'alimentation des poulets de chair au Sud-Kivu, République Démocratique du Congo. *Tropicultura*, 20(1): 10–16.
Murray, S.S., Schoeninger, M.J., Bunn, H.T., Pickering, T.R. & Marlett, J.A. 2001. Nutritional composition of some wild plant foods and honey used by Hadza foragers of Tanzania. *Journal of Food Composition and Analysis*, 14: 3–13.
Mustafa, N.E.M., Mariod, A.A., & Matthäus, B. 2008. Antibacterial activity of *Aspongopus viduatus* (Melon bug) oil. *Journal of Food Safety*, 28: 577–586.
Muyay, T. 1981. *Les insectes comme aliments de l'homme: Serie II, Vol. 69*. Democratic Republic of the Congo, Ceeba Publications.
Muzzarelli, R.A.A. 2010. Chitins and chitosans as immunoadjuvants and non-allergenic drug carriers. *Marine Drugs*, 8(2): 292–312.
Muzzarelli, R.A.A., Terbojevich, M., Muzzarelli, C., Miliani, M. & Francescangeli, O. 2001. Partial depolymerization of chitosan with the aid of papain. *In* R.A.A. Muzzarelli, ed. *Chitin Enzymology*, pp. 405–414. Italy, Atec.
Myers, N., Mittermeier, R.A., Mittermeier, C.G., De Fonseca, G.A.B. & Kent, J. 2000. Biodiversity hotspots for conservation priorities. *Nature*, 403: 853–858.
Neuenschwander, P., Sinsin, B. & Goergen, G., eds. 2011. *Protection de la nature en Afrique de l'Ouest: une liste rouge pour le Bénin*. Nature conservation in West Africa: red list for Benin. Ibadan, Nigeria, International Institute of Tropical Agriculture.
N'Gasse, G. 2004. Contribution des insectes de la forêt à la sécurité alimentaire. *Produits Forestiers Non Ligneux: Document de Travail 1*. Rome, FAO.
Nakagaki, B.J. & De Foliart, G.R. 1991. Comparison of diets for mass-rearing *Acheta domesticus* (Orthoptera: Gryllidae) as a novelty food, and comparison of food conversion efficiency with values reported for livestock. *Journal of Economic Entomology*, 84(3): 891–896.
Naughton, J.M., Odea, K. & Sinclair, A.J. 1986. Animal foods in traditional Australian aboriginal diets: polyunsaturated and low in fat. *Lipids*, 21(11): 684–690.

Neely, G.G., Keene, A.C., Duchek, P., Chang, E.C., Wang, Q.P., Aksoy, Y.A., Rosenzweig, M., Costigan, M., Woolf, C.J., Garrity, P.A. & Penninger, J.M. 2011. TrpA1 regulates thermal nociception in Drosophila. *Plos One*, 6(8): e24343.

Newton, G.L., Booram, C.V., Barker, R.W. & Hale, O.M. 1977. Dried *Hermetia illucens* larvae meal as supplement for swine. *J. Anim Sci.*, 44: 395–400.

Newton, L., Sheppard, C., Watson, D.W. & Burtle, G. 2005. *Using the black soldier fly,* Hermetia illucens, *as a value-added tool for the management of swine manure.* North Carolina, North Carolina State University. (available at www.cals.ncsu.edu/waste_mgt/smithfield_projects/phase2report05/cd,web%20files/A2.pdf)

Nkouka, E. 1987. Les insectes comestibles dans les societes d'Afrique Centrale. *Revue Scientifique et Culturelle du CICIBAASC LEIDEN*, 6(1): 171–178.

Nonaka, K. 1996. Ethnoentomology of the Central Kalahari San. *African Study Monographs*, 22: 29–46.

Nonaka, K. 2007. *Hebo yellow jackets: from the fields to the dinner table: a delightful culinary experience.* Tokyo, Tamasaya.

Nonaka, K. 2009. Feasting on insects. (Special issue: trends on the edible insects in Korea and Abroad.). *Entomological Research*, 39(5): 304–312.

Nonaka, K. 2010. Cultural and commercial roles of edible wasps in Japan. *In* P.B. Durst, D.V. Johnson, R.L. Leslie. & K. Shono, eds. *Forest insects as food: humans bite back, proceedings of a workshop on Asia-Pacific resources and their potential for development.* pp. 123–130. Bangkok, FAO Regional Office for Asia and the Pacific.

Nonaka, K., Sivilay, S. & Boulidam, S. 2008. *The biodiversity of insects in Vientiane.* Nara, Japan, National Agriculture and Forestry Institute and Research Institute for Hamanity and Nature.

Nordic Food Lab. 2012. Finding the deliciousness of insects. (available at http://nordicfoodlab.org/blog/2012/07/mad-2-finding-the-deliciousness-of-insects). Accessed November, 2012.

O'Dea, K., Jewell, P.A., Whiten, A., Altmann, S.A., Strickland, S.S. & Oftedal, O.T. 1991. Traditional diet and food preferences of Australian Aboriginal hunter-gatherers. *Philosophical Transactions of the Royal Society of London Series B.*, 334: 233–241.

Offenberg, J. 2011. *Oecophylla smaragdina* food conversion efficiency: prospects for ant farming. *Journal of Applied Entomology*, 135(8): 575–581.

Offenberg, J. & Wiwatwitaya, D. 2009a. Sustainable weaver ant (*Oecophylla smaragdina*) farming: harvest yields and effects on worker ant density. *Asian Myrmecology*, 3: 55–62.

Offenberg, J. & Wiwatwitaya, D. 2009b. Weaver ants convert pest insects into food: prospects for the rural poor. Paper presented at the International Conference on Research, Food Security, Natural Resource Management and Rural Development, University of Hamburg, Germany, 6–8 October 2009.

Ogutu, M.A. 1986. Sedentary hunting and gathering among the Tugen of Baringo District in Kenya. *Sprache und Geschichte in Afrika*, 7(2): 323–338.

Ohura, M. 2003. Development of an automated warehouse type silkworm rearing system for the production of useful materials. *Journal of Insect Biotechnology and Sericology*, 72(3): 163–169.

Oonincx, D.G.A.B. & van der Poel, A.F.B. 2011. Effects of diet on the chemical composition of migratory locusts (*Locusta migratoria*). *Zoo Biology*, 30: 9–16.

Oonincx, D.G.A.B. & de Boer, I.J.M. 2012. Environmental impact of the production of mealworms as a protein source for humans: a life cycle assessment. *PLoS ONE*, 7(12): e51145.

Oonincx, D.G.A.B., van Itterbeeck, J., Heetkamp, M.J.W., van den Brand, H., van Loon, J. & van Huis, A. 2010. An exploration on greenhouse gas and ammonia production by insect species suitable for animal or human consumption. *Plos One*, 5(12): e14445.

Osmaston, H.A. 1951. The termite and its uses for food. *Uganda Journal (Kampala)*, 15: 8–83.

Oso, B. 1977. Mushrooms in Yoruba mythology and medicinal practices. *Economic Botany*, 31: 367–371.

Oudhia, P. 2002. Traditional medicinal knowledge about red ant *Oecophylla smaragdina* (Fab.) (Hymenoptera: Formicidae) in Chhattisgarh, India. *Insect Environment*, 8(3): 114–115.

Owen, D.F. 1973. *Man's environmental predicament: an introduction to human ecology in tropical Africa*. Oxford, Oxford University Press.

Pagezy, H. 1975. Les interrelations homme faune de la foret du Zaire. *l'Homme et l'Animal, Premier Colloque d'Ethnozoologie*, pp.63–68. Paris, Institut International d'Ethnosciences.

Panizzi, A.R. 1997. Wild hosts of Pentatomids: Ecological significance and role in their pest status on crops. *Annual Reviews of Entomology*, 42: 99–122.

Paoletti, M.G. ed. 2005. *Ecological implications of minilivestock; role of rodents, frogs, snails, and insects for sustainable development*. New Hampshire, USA, Science Publishers.

Paoletti, M.G., Buscardo, E., Vanderjagt, D.J., Pastuszyn, A., Pizzoferrato, L., Huang, Y.S., Chuang, L.T., Glew, R.H., Millson, M. & Cerda, H. 2003. Nutrient content of termites (*Syntermes* soldiers) consumed by Makiritare Amerindians of the Alto Orinoco of Venezuela. *Ecology of Food and Nutrition*, 42(2): 177–191.

Paoletti, M.G. & Dufour, D.L. 2005. Edible invertebrates among Amazonian Indians: a critical review of disappearing knowledge. *In* M.G. Paoletti, ed. *Ecological implications of minilivestock; role of rodents, frogs, snails, and insects for sustainable development*, pp. 293–342. New Hampshire, Science Publishers.

Paoletti, M.G., Dufour, D.L., Cerda, H., Torres, F., Pizzoferrato, L. & Pimentel, D. 2000. The importance of leaf- and litter-feeding invertebrates as sources of animal protein for the Amazonian Amerindians. *Proceedings of the Royal Society of London*, 267: 1459, 2247–2252.

Paoletti, M.G., Norberto, L., Damini, R. & Musumeci, S. 2007. Human gastric juice contains chitinase that can degrade chitin. *Annals of Nutrition and Metabolism*, 51(3): 244–251.

Parent, G. & Thoen, D. 1977. Food value of edible mushrooms from upper Shaba region. *Economic Botany*, 31: 436–445.

Parsons, J.R. 2010. The pastoral niche in Pre-Hispanic Mesoamerica. *In* J.E. Staller and M.D. Carrasco. *Pre-Columbian foodways: interdisciplinary approaches to food, culture and markets in ancient Mesoamerica*, pp. 109–136. New York, Springer.

Pearce, M.J. 1997. *Termites: biology and pest management*. Wallingford. CAB International.

Pemberton, R.W. 1988. The use of the giant waterbug, *Lethocerus indicus* (Hemiptera: Belostomatidae), as human food in California. *Pan-Pacific Entomology*, 64(1): 81–82.

Pemberton, R.W. 1994. The revival of rice-field grasshoppers as human food in South Korea. *Pan-Pacific Entomologist*, 70(4): 323–327.

Peng, R.K. & Christian, K. 2005. Integrated pest management in mango orchards in the Northern Territory Australia, using the weaver ant, *Oecophylla smaragdina* (Hymenoptera: Formicidae) as a key element. *International Journal of Pest Management*, 51(2): 149–155.

Peng, R.K., Christian, K. & Gibb, K. 2004. Implementing ant technology in commercial cashew plantations and continuation of transplanted green ant colony monitoring. Canberra, Rural Industries Research and Development Corporation.

Pérez-Expósito, A.B. & Klein, B.P. 2009. Impact of fortified blended food aid products on nutritional status of infants and young children in developing countries. *Nutrition Review*, 67(12): 706–718.

Perez, M.R. 1995. *A conceptual framework for CIFOR's research on non-wood forest products*. CIFOR Working Paper, 6. Bogor, CIFOR.

Phillips, J.K. & Burkholder, W.E. 1995. Allergies related to food insect production and consumption. *The Food Insects Newsletter*, 8(2): 1, 2–4.

Pimentel, D. 1991. Ethanol fuels: Energy security, economics and the environment. *Agri. and Envir. Ethics*, 4(1): 1–13.

Pimentel, D. & Pimentel, M. 2003. Sustainability of meat-based and plant-based diets and the environment. *Am. J. Clin. Nutr.*, 78: 660S–663S.

Pimentel, D., Berger, B., Filiberto, D., Newton, M., Wolfe, B., Karabinakis, E., Clark, S., Poon, E., Abbett, E. & Nandagopal, S. 2004. Water resources: agricultural and environmental issues. *BioScience*, 54: 909–918.

Pimentel, D., Dritschilo, W., Krummel, J. & Kutzman, J. 1975. Energy and land constraints in food protein production. *Science*, 190 (4216): 754–761.

Pliner, P. & Salvy, S.J. 2006. Food neophobia in humans. *In* R. Shepherd & M. Raats, eds. *The psychology of food choice*, pp. 75–92. Oxfordshire, CABI Publishing.

Politis, G.G. 1996. Moving to produce: Nukak mobility and settlement patterns in Amazonia. *World Archaeology*, 27(3): 492–511.

Portes, E., Gardrat, C., Castellan, A. & Coma, V. 2009. Environmentally friendly films based on chitosan and tetrahydrocurcuminoid derivatives exhibiting antibacterial and antioxidative properties. *Carbohydrate Polymers*, 76(4): 578–584.

Raina, S.K., Kioko, E.N., Gordon, I. & Nyandiga, C. 2009. *Commercial insects and forest conservation: improving forest conservation and community livelihoods through income generation from commercial insects in three Kenyan forests*. Nairobi, Science Press.

Ramandey, E. & van Mastrigt, H. 2010. Edible insects in Papua, Indonesia: from delicious snack to basic need. *In* P.B. Durst, D.V. Johnson, R.L. Leslie. & K. Shono, eds. *Forest insects as food: humans bite back, proceedings of a workshop on Asia-Pacific resources and their potential for development*. pp. 105–114. Bangkok, FAO Regional Office for Asia and the Pacific.

Ramos Elorduy, J. 1984. Los insectos como un recurso actual et potencial. *In* T.R. Trujillo, ed. *Seminario sobre la alimentacion en Mexico*, pp. 126–139. Mexico, Instituto de Geografia, National Autonomous University of Mexico.

Ramos Elorduy, J. 1990. Edible insects: barbarism or solution to the hunger problem? *In* D.A. Posey & W.L. Overal, eds. *Ethnobiology: implications and applications. Proceedings of the First International Congress of Ethnobiology*, pp. 151–158. Bélem, Brazil, Museu Paraense Emílio Goeldi.

Ramos Elorduy, J. 1993. Food production and nutrional value of wild and semi-domesticated species: background. *In* C.M. Hladik, A. Hladik, O.F. Linares, H. Pagezy, A. Semple & M. Hadley, eds. *Tropical forests, people and food: biocultural interactions and applications to development*. Man and the Biosphere Series Volume 13. Paris, United Nations Educational, Scientific and Cultural Organization.

Ramos Elorduy, J. 1997. The importance of edible insects in the nutrition and economy of people of the rural areas of Mexico. *Ecology of Food and Nutrition*, 36: 347–366.

Ramos Elorduy, J. 2005. Insects: a hopeful food source. *In* M.G. Paoletti, ed. *Ecological implications of minilivestock; role of rodents, frogs, snails, and insects for sustainable development*, pp. 263–291. New Hampshire, Science Publishers.

Ramos Elorduy, J. 2006. Threatened edible insects in Hidalgo, Mexico and some measures to preserve them. *Journal of Ethnobiology and Ethnomedicine*, 2(51): 1–10.

Ramos Elorduy, J. 2009. Anthropo-entomophagy: cultures, evolution and sustainability. (Special issue: trends on the edible insects in Korea and abroad). *Entomological Research*, 39: 5: 271–288.

Ramos Elorduy, J., Carbajal Valdés, L.A. & Pino, J.M. 2012. Socio-economic and cultural aspects associated with handling grasshopper germplasm in traditional markets of Cuautla, Morelos, Mexico. *J. Hum. Ecol.*, 40(1): 85–94.

Ramos Elorduy, J., Costa-Neto, E.M., Pino, J.M., Cuevas Correa, M.S., Garcia-Figueroa, J. & Zetina, D.H. 2007. Conocimiento de la entomofauna útil en la Purísima Palmar de Bravo, Puebla, México. *Biotemas*, 20(2): 121–134.

Ramos Elorduy, J., Gonzalez, E.A., Hernandez, A.R. & Pino, J.M. 2002. Use of *Tenebrio molitor* (Coleoptera: Tenebrionidae) to recycle organic wastes and as feed for broiler chickens. *Journal of Economic Entomology*, 95(1): 214–220.

Ramos Elorduy, J. & Pino, J.M. 1989. Los insectos comestibles en el México antiguo (estudio etnoentomológico). Mexico, A.G.T. Editor México.

Ramos Elorduy, J. & Pino, J.M. 2002. Edible insects of Chiapas, Mexico. *Ecology of Food and Nutrition*, 41(4): 271–299.

Ramos Elorduy, J. & Pino, J.M. 2003. Enfermedades tratadas con insectos en el continente americano. *Entomología Mexicana*, 2: 604–611.

Ramos Elorduy, J. & Pino, J.M. 2006. Algunos ejemplos de aprovechamiento comercial de varios insectos comestibles y medicinales. *Entomología Mexicana*, 1: 524–533.

Ramos Elorduy, J., Pino, J.M. & Martínez, V.H.C. 2008. Una vista a la biodiversidad de la antropoentomofagía mundial. *Entomología Mexicana*, 7: 308–313.

Ramos Elorduy, J., Pino, J. M. & Martínez, V.H.C. 2009. Edible aquatic Coleoptera of the world with an emphasis on Mexico. *Journal of Ethnobiology and Ethnomedicine*, 5(11).

Ramos Elorduy, J., Pino, J.M., Prado, E.E., Perez, M.A., Otero, J.L. & de Guevara, O.L. 1997. Nutritional value of edible insects from the state of Oaxaca, Mexico. *Journal of Food Composition and Analysis*, 10: 142–157.

Ramos Elorduy, J., Pino, J.M., Vázquez, A.I., Landero, I., Oliva-Rivera, H. & Martinez, V.H.C. 2011. Edible Lepidoptera in Mexico: Geographic distribution, ethnicity, economic and nutritional importance for rural people. *Journal of Ethnobiology and Ethnomedicine*, 7(2): 1–22.

Rastogi, N. 2011. Provisioning services from ants: food and pharmaceuticals. *Asian Myrmecology*, 4: 103–120.

Ravindran, V. & Blair, R. 1993. Feed resources for poultry production in Asia and the Pacific. *World's Poulty Science Journal*, 49: 219–235.

Reese, G., Ayuso, R. & Lehrer, S.B. 1999. Tropomyosin: An invertebrate pan-allergen. *International Archives of Allergy and Immunology*, 119(4): 247–258.

Roberge, J.M. & Angelstam, A.P. 2004. Usefulness of the umbrella species concept as a conservation tool. *Conservation Biology*, 18(1): 76–85.

Roberts, C. 1998. Long-term costs of the mopane worm harvest. *Oryx*, 32(1): 6–8.

Rohde, T.S. 2012. Jysk præst spiste græshoppe under gudstjeneste: Kollega siger op i protest. In *BT*, posted on 12 September 2012. (available at www.bt.dk/utroligt-men-sandt/jysk-praest-spiste-graeshoppe-under-gudstjeneste-kollega-siger-op-i-pro). Accessed November 2012.

Roos, N., Nurhasan, M., Thang, B., Skau, J., Wieringa, F., Khov, K., Friis, H., Michaelsen, K.F. & Chamnan, C. 2010. WinFood Cambodia: improving child nutrition through improved utilization of local food. Poster for The WinFood Project, Denmark, Department of Human Nutrition, University of Copenhagen.

Roulon-Doko, P. 1998. Chasse, cueillette et cultures chez les Gbaya de Centrafrique. Paris, L'Harmattan.

Rozin, P. & Fallon, A.E. 1987. A perspective on disgust. *Psychological Review*, 94(1): 23–41.

Rozin, P. & Vollmecke, T.A. 1986. Food likes and dislikes. *Annual Review Nutrition*, 6: 433–456.

Rumpold, B.A. & Schlüter, O.K. 2013. Nutritional composition and safety aspects of edible insects. *Molecular Nutrition and Food Research*, 57(3) (DOI 10.1002/mnfr.201200735).

Rutaisire, J. 2007. Analysis of feeds and fertilizers for sustainable aquaculture development in Uganda. *In* M.R. Hasan, T. Hecht, S.S. De Silva & A.G.J. Tacon, eds. *Study and analysis of feeds and fertilizers for sustainable aquaculture development*, pp. 471–488. Rome, FAO.

Ryan, L.G. 1996. *Insect musicians & cricket champions: a cultural history of singing insects in China and Japan*. San Francisco, China Books and Periodicals, Inc.

Ryu, K.S., Lee, H.S., Kim, K.Y., Kim, M.J., Kang, P.D., Chun, S.N., Lee, S.H. & Lee, M.L. 2012. Anti-diabetic effects of the silkworm (*Bombyx mori*) extracts in the db/db mice. *Planta Med.*, 78: 458.

Sachs, J. 2010. Rethinking macroeconomics: knitting together global society. *The Broker*, 10: 1–3.

Saeed, T., Dagga, F.A. & Saraf, M. 1993. Analysis of residual pesticides present in edible locusts captured in Kuwait. *Arab Gulf Journal of Scientific Research*, 11(1): 1–5.

Sahagun, B. 1557. *Historia de las cosas de Nueva España (1905)*, 6: 2. Madrid, Hauser y Menet.

Samways, M.J. 2007. Insect conservation: a synthetic management approach. *Ann. Rev. Entomol*, 52: 465–487.

Sanderson, M.G. 1996. Biomass of termites and their emissions of methane and carbon dioxide: a global database. *Global Biogeochemical Cycles*, 10(4): 543–557.

Sankar, S. 2001. *Environmental economics*. Chennai, India, Margham Publications.

Santos Oliveira, J. F., Passos de Carvalho, J., Bruno de Sousa, R.F.X. & Madalena Simao, M. 1976. The nutritional value of four species of insects consumed in Angola. *Ecology of Food and Nutrition*, 5: 91–97.

Schabel, H. 2006. Forest-based insect industries. *In* H. Schabel, ed. *Forest entomology in East Africa: forest insects of Tanzania*, pp. 247–294.

Schneider, J. 1844. Maikäfersuppen, ein vortreffliches und kräftiges Nahrungsmittel. *Magazin für die Staatsarzneikunde*, 3: 403–405.

Schneider, J.C. ed. 2009. *Principles and procedures for rearing high quality insects*. USA, Mississippi State University.

Scholtz, C.H. 1984. *Useful insects*. Pretoria, De Jager-HAUM Publishers.

Schroeckenstein, D.C., Meier-Davis, S. & Bush, R.K. 1990. Occupational sensitivity to *Tenebrio molitor* Linnaeus (yellow mealworm). *The Journal of Allergy and Clinical Immunology*, 86(2): 182–188.

Schroeckenstein, D.C., Meier-Davis, S., Graziano, F.M., Falomo, A. & Bush, R.K. 1988. Occupational sensitivity to *Alphitobius diaperinus* (Panzer) (lesser mealworm). *The Journal of Allergy and Clinical Immunology*, 82(6): 1081–1088.

Sekhwela, M.D.B. 1988. The nutritive value of mophane bread: mophane insect secretion (Maphote or Maboti). *Botswana Notes and Records*, 20: 151–153.

Senti, G., Lundberg, M. & Wüthrich, B. 2000. Asthma caused by a pet bat. *Allergy*, 55(4): 406–407.

Settle, W.H., Ariawan, E.T., Astuti, W., Cahyana, A.L., Hakim, D., Hindayana, A.S. & Pajamingsih, L. 1996. Managing tropical rice pests through conservation of generalist natural enemies and alternative prey. *Ecology*, 77: 1975–1988.

Shear, W.A. & Kukalová-Peck, J. 1990. The ecology of Paleozoic terrestrial arthropods: the fossil evidence. *Canadian Journal of Zoology*, 68(9): 1807–1834.

Shen, L., Li, D., Feng, F. & Ren, Y. 2006. Nutritional composition of *Polyrhachis vicina* Roger (edible Chinese black ant). *Songklanakarin Journal of Science and Technology*, 28(1): 107–114.

Sheppard, D.C. 1983. House fly and lesser fly control utilizing the black soldier fly in manure management systems for caged laying hens. *Environ. Entomol.*, 12: 1439–1442.

Sheppard, D.C., Newton, G.L. & Burtle, G. 2008. Black soldier fly prepupae: a compelling alternative to fish meal and fish oil. A public comment prepared in response to a request by the National Marine Fisheries Service to gather information for the NOAA-USDA Alternative Feeds Initiative. Public comment on alternative feeds for aquaculture received by NOAA 15 November 2007 through 29 February 2008.

Sheppard, D.C., Newton, G.L., Thompson, S.A. & Savage, S. 1994. A value added manure management system using the black soldier fly. *Bioresource Technology*, 50(3): 275–279.

Sherman, R.A. & Wyle, F.A. 1996. Low-cost, low maintenance rearing of maggots in hospitals, clinics, and schools. *Am. J. Trop. Med. Hyg.*, 54: 38–41.

Silow, C.A. 1983. Notes on Ngangela and Nkoya ethnozoology. Ants and termites. *Etnologiska Studier*, 36: 177.

Simberloff, D. 1998. Flagships, umbrellas, and keystones: is single-species management passé in the landscape era? *Biological Conservation*, 83(3): 247–257.

Singtripop, T., Wanichacheewa, S. & Sakurai, S. 2000. Juvenile hormone-mediated termination of larval diapause in the bamboo borer, *Omphisa fuscidentalis*. *Insect Biochemistry and Molecular Biology* 30: 847–854.

Siracusa, A., Marcucci, F., Spinozzi, F., Marabini, A., Pettinari, L., Pace, M.L. & Tacconi, C. 2003. Prevalence of occupational allergy due to live fish bait. *Clinical and Experimental Allergy*, 33(4): 507–510.

Sirimungkararat, S., Saksirirat, W., Nopparat, T. & Natongkham, A. 2010. Edible products from eri and mulberry silkworms in Thailand. *In* P.B. Durst, D.V. Johnson, R.L. Leslie. & K. Shono, eds. *Forest insects as food: humans bite back, proceedings of a workshop on Asia-Pacific resources and their potential for development*. pp. 189–200. Bangkok, FAO, Regional Office for Asia and the Pacific.

Siripanthong, S., Teerapantuwat, S., Prugsnusak, W., DSuputtamongkol, Y., Viriyasithavat, P., Chaowagul, W., Dance, D.A.B. & White, N.J. 1991. Corneal ulcer cause by *Pseudomonas pseudomallei*: Report of three cases. *Reviews of Infectious Diseases*, 13(2): 335–337.

Skelton, G.S. & Matanganyidze, C. 1981. Detection by quantitative assay of various enzymes in the edible mushroom *Termitomyces microcarpus*. *Bull. Soc. Bot. Fr. 128, Lettres Botanique*, 3: 143–149.

Slingenbergh, J., Gilbert, M., de Balogh, K. & Wint, W. 2004. Ecological sources of zoonotic diseases. *Rev. Sci. Tech. Off. Int. Epiz.*, 23(2): 467–484.

Smil, V. 2002. Worldwide transformation of diets, burdens of meat production and opportunities for novel food proteins. *Enzyme and Microbial Technology*, 30: 305–311.

Smith, A.B.T. & Paucar, C.A. 2000. Taxonomic review of *Platycoelia lutescens* (Scarabaeidae: Rutelinae: Anoplognathini) and a description of its use as food by the people of the Ecuadorian highlands. *Annals of the Entomological Society of America*, 93(3): 408–414.

Sogbesan, A. & Ugwumba, A. 2008. Nutritional evaluation of termite (*Macrotermes subhyalinus*) meal as animal protein supplements in the diets of Heterobranchus longifilis. *Turkish Journal of Fisheries and Aquatic Sciences*, 8: 149–157.

Sribandit, W., Wiwatwitaya, D., Suksard, S. & Offenberg, J. 2008. The importance of weaver ant (*Oecophylla smaragdina* Fabricus) harvest to a local community in Northeastern Thailand. *Asian Myrmecology*, 2: 129–138.

St-Hilaire, S., Cranfill, K., Mcguire, M.A., Mosley, E.E., Tomberlin, J.K., Newton, L., Sealey, W., Sheppard, C. & Irving, S. 2007. Fish offal recycling by the Black Soldier Fly produces a foodstuff high in omega-3 fatty acids. *Journal of the World Aquaculture Society*, 38(2): 309–313.

Stack, J., Dorward, A., Gondo, T., Frost, P., Taylor, F. & Kurebgaseka, N. 2003. Mopane worm utilisation and rural livelihoods in Southern Africa. Paper presented at the International Conference on Rural Livelihoods, Forests and Biodiversity, Bonn, Germany, 19–23 May 2003.

Steinfeld, H., Gerber, P., Wassenaar, T., Castel, V., Rosales, M. & de Haan, C, eds. 2006. *Livestock's long shadow: environmental issues and options*. Rome, FAO.

Sunderland, T.C.H., Ndoye, O. & Harrison-Sanchez, S. 2011. Non-timber forest products and conservation: what prospects? *In* S. Shackleton, Ch. Shackleton & P. Shanley, eds. *Non-timber forest products in the global context*, pp. 209–224. Heidelberg, Germany, Springer.

Sweet, C. 2011. Sausalito insect supper promises finely sourced bugs. *The Sanfrancisco Gate*, posted on 23 September 2011. (available at http://insidescoopsf.sfgate.com/blog/2011/09/23/sausalito-insect-supper-promises-finely-sourced-bugs/).

Tabuna, H. 2000. *Evaluation des échanges des produits forestiers non ligneux entre l'Afrique subsaharienne et l'Europe*. Accra, FAO Regional Office for Africa.

Tacon, A.G.J. & Metian, M. 2008. Global overview on the use of fish meal and fish oil in industrially compounded aquafeeds: Trends and future prospects. *Aquaculture*, 285: 146–158.

Takeda, J. 1990. The dietary repertory of the Ngandu people of the tropical rain forest: an ecological and anthropological study of the subsistence activities and food procurement technology of a slash-and-burn agriculturist in the Zaire river basin. *African Study Monographs. Supplementary Issue*, 11: 1–75.

Takeda, J. & Sato, H. 1993. Multiple subsistence strategies and protein resources of horticulturists in the Zaire basin: the Nganda and the Boyela. *In* C.M. Hladik, A. Hladik, O.F. Linares, H. Pagezy, A. Semple & M. Hadley, eds. *Tropical forests, people and food: biocultural interactions and applications to development.* Man and the Biosphere Series, 13. Paris, United Nations Educational, Scientific and Cultural Organization.

Taylor, R.L. 1975. *Butterflies in my stomach: insects in human nutrition.* Santa Barbara, USA, Woodbridge Press Publishing Company.

Teffo, L.S. 2006. *Nutritional and medicinal value of the edible stinkbug,* Encosternum delegorguei *Spinola consumed in the Limpopo Province of South Africa and its host plant* Dodoneae viscosa *Jacq. var. angustifolia*. Pretoria, University of Pretoria. (Phd thesis).

Téguia, A., Mpoame, M. & Okourou, M.J.A. 2002. The production performance of broiler birds as affected by the replacement of fish meal by maggot meal in the starter and finisher diets. *Tropicultura*, 20(4): 187–192.

Tempelado, L. 2012. Insect connoisseurs ask: Got any good recipes? *Asahi Shimbun*, online edition, posted on 12 December 2012. (available at http://ajw.asahi.com/article/cool_japan/cooking/AJ201212120003).

Thorne, P.S. 2007. Environmental health impacts of concentrated animal feeding operations: anticipating hazards: searching for solutions. *Environ. Health Perspect.*, 115: 296–297.

Tieguhong, J.C., Ndoye, O., Vantomme, P., Grouwels, S., Zwolinski, J. & Masuch, J. 2009. Coping with crisis in Central Africa: enhanced role for non-wood forest products. *Unasylva*, 233(60): 49–54.

Tihon, L. 1946. A propos des termites au point de vue alimentaire. *Bull. Agric. du Congo Belge*, 37: 865–868.

Tilman, D., Cassman, K.G., Matson, P.A., Naylor, R. & Polasky, S. 2002. Agricultural sustainability and intensive production practices. *Nature*, 418: 671–677.

Tiu, L.G. 2012. Enhancing Sustainability of Freshwater Prawn Production in Ohio. *Ohio State University South Centers Newsletter*, Fall 2012.

Toledo, A. & Burlingame, B. 2006. Biodiversity and nutrition: a common path toward global food security and sustainable development. *Journal of Food Composition and Analysis*, 19: 477–483.

Tomberlin, J.K. & Sheppard, D.C. 2001. Lekking behavior of the black soldier fly (Diptera: Stratiomyidae). *Florida Entomologist*, 84(4): 729–730.

Tommaseo Ponzetta, M. & Paoletti, M.G. 1997. Insects as food of the Irian Jaya populations. *Ecology of Food and Nutrition*, 36: 321–346.

Toms, R. & Thagwana, M. 2005. On the trail of missing mopane worms. *Science in Africa*. (available at www.scienceinafrica.co.za/2005/january/mopane.htm).

Toms, R. & Thagwana, M. 2003. Eat your bugs! *Science in Africa*. (available at www.scienceinafrica.co.za/2005/january/mopane.htm).

Tong, L., Yu, X. & Lui, H. 2011. Insect food for astronauts: gas exchange in silkworms fed on mulberry and lettuce and the nutritional value of these insects for human consumption during deep space flights. *Bulletin of Entomological Research*, 101: 613–622.

Torres, A.E. 2008. *Conociendo la cadena productive de tuna y cochinilla en Ayacucho.* Ayachucho, Solid Peru.

Townsend, P.K. 1973. Sago production in a New Guinea economy. *Human Ecology*, 2: 217–236.

Triplehorn, C.A. & Johnson, N.F., eds. 2004. *Borror and DeLong's introduction to the study of insects.* St. Paul, USA, Brooks Cole.

Turner, J.S. & Soar, R.C. 2008. *Beyond biomimicry: what termites can tell us about realizing the living building.* Presentation to the First International Conference on Industrialized, Intelligent Construction (I3CON), 14–16 May, 2008, Loughborough University, England.

Twine, W., Moshe, D., Netshiluvhi, T. & Siphugu, V. 2003. Consumption and direct-use values of savanna bio-resources used by rural households in Mametja, a semi-arid area of Limpopo province, South Africa. *South African Journal of Science*, 99(9–10): 467–473.

Umesh, K.B., Akshara, M., Shripad, B., Harish, K.K. & Srinivasan, S.M. 2009. Performance analysis of production and trade of Indian silk under WTO regime. Paper presented at the International Association of Agricultural Economists Conference, 16–22 August, Beijing, China.

UN. 2012. *World urbanization prospects, the 2011 revision.* New York, USA.

UNESCO. 2005. *United Nations decade of education for sustainable development (2005–2014): draft international implementation scheme.* Paris, United Nations Educational, Scientific and Cultural Organization.

Urs, K.C.D. & Hopkins, T.L. 1973a. Effect of moisture on growth rate and development of two strains of *Tenebrio molitor* L. (Coleoptera, Tenebrionidae). *Journal of Stored Products Research*, 8: 291–297.

Urs, K.C.D. & Hopkins, T.L. 1973b. Effect of moisture on the lipid content and composition of two strains of *Tenebrio molitor* L. (Coleoptera, Tenebrionidae). *Journal of Stored Products Research*, 8(4): 299–305.

USDA. 2012. USDA National Nutrient Database for Standard Reference, Release 25. (Latest update September 2012). Accessed December 2012. (available at www.ars.usda.gov/ba/bhnrc/ndl).

USFDA. 2009. Color additives: FDA's regulatory process and historical perspectives. (available at www.fda.gov/ForIndustry/ColorAdditives/RegulatoryProcessHistoricalPerspectives/default.htm). Accessed November 2012.

van der Meer, K. 2004. Exclusion of small-scale farmers from coordinated supply chains: Market failure, policy failure or just economies of scale? Paper presented at the workshop "Is there a place for smallholder producers in coordinated supply chains?", 8 December 2004, Washington, D.C., USA, World Bank.

van Hall, M.A.L., Dierikx, C.M., Cohen, S.J., Voets, G.M., van den Munckhof, M.P., van Essen-Zandbergen, A., Platteel, T., Fluit, A.C., van de Sande-Bruinsma, N., Scharinga, J. Bonten, M.J.M. & Mevius, D.J. 2011. Dutch patients, retail chicken meat and poultry share the same ESBL genes, plasmids and strains. *Clinical Microbiology and Infection*, 17(6): 873–880.

van Huis, A. 1996. The traditional use of arthropods in Sub-Saharan Africa. *Proceedings of the section Experimental and Applied Entomology of the Netherlands Entomological Society (N.E.V.)*, 7: 3–20.

van Huis, A. 2003a. Medical and stimulating properties ascribed to arthropods and their products in sub-Saharan Africa. *In* É. Motte-Florac & J.M.C. Thomas, eds. *Insects in oral literature and traditions*, pp. 367–382. Ethnosciences: 11. Société d'études linguistiques et anthropologiques de France (series): 407. Paris, Peeters.

van Huis, A. 2003b. Insects as food in sub-Saharan Africa. *Insect Science and its Application*, 23(3): 163–185.

van Huis, A. 2013. Potential of insects as food and feed in assuring food security. *Annual Review of Entomology*, 58(1): 563–583.
van Huis, A. 2005. Insects eaten in Africa (Coleoptera, Hymenoptera, Diptera, Heteroptera, Homoptera). *In* M.G. Paoletti, ed. *Ecological implications of minilivestock*, pp. 231–244. New Hampshire, USA, Science Publishers.
van Huis, A., van Gurp, H. & Dicke, M. 2012. *Het insectenkookboek*. Amsterdam, the Netherlands, Atlas.
van Itterbeeck, J. 2008. Entomophagy and the West: barriers and possibilities, ecological advantages and ethical desirability. Wageningen, Wageningen University. (Phd thesis).
van Itterbeeck, J. & van Huis, A. 2012. Environmental manipulation for edible insect procurement: a historical perspective. *Journal of Ethnobiology and Ethnomedicine*, 8(3): 1–19.
van Lenteren, J.C. 2006. Ecosystem services to biological control of pests: why are they ignored? *Proc. Neth. Entomol. Soc. Meet.*, 17: 103–111.
van Mele, P. 2008. A historical review of research on the weaver ant Oecophylla in biological control. *Agricultural and Forest Entomology*, 10: 13–22.
Vane-Wright, R.I. 1991. Why not eat insects? *Bulletin of Entomological Research*, 81: 1–4.
Vantomme, P., Gazza, S. & Lescuyer, G. 2010. *Produits forestiers non ligneux*, Vol. 304. Paris, Centre de coopération internationale en recherche agronomique pour le développement.
Vantomme, P., Göhler, D. & N'Deckere-Ziangba, F. 2004. Contribution of forest insects to food security and forest conservation: The example of caterpillars in Central Africa. *Odi Wildlife Policy Briefing*, 3.
Vega, F. & Kaya, H. 2012. *Insect Pathology*. London, Academic Press.
Veldkamp, T., G. van Duinkerken, A. van Huis, C.M.M. Lakemond, E. and Ottevanger, E., and M.A.J.S van Boekel, 2012. *Insects as a sustainable feed ingredient in pig and poultry diets. A feasibility study*. Wageningen UR Livestock Research, Report 638.
Vepari, C. & Kaplan, D. 2007. Silk as a biomaterial. *Progress in Polymer Science*, 32(8-9): 99–107.
Vernon, L.L. & Berenbaum, H. 2004. A naturalistic examination of positive expectations, time course, and disgust in the origins and reduction of spider and insect distress. *Anxiety Disorders*, 18: 707–718.
Vijver, M., Jager, T., Posthuma, L. & Peijnenburg, W. 2003. Metal uptake from soils and soil-sediment mixtures by larvae of *Tenebrio molitor* (L.) (Coleoptera). *Ecotoxicology and Environmental Safety*, 54(3): 277–289.
Villanueva, G.R., Roubik, D.W. & Colli-Ucan, W. 2005. Extinction of *Melipona beecheii* and traditional beekeeping in the Yucatan peninsula. *Bee World*, 86(2): 35–41.
Wang, D., Bai, Y.T., Li, J.H. & Zhang, C.X. 2004. Nutritional value of the field cricket (*Gryllus testaceus* Walker). *Journal of Entomologia Sinica*, 11: 275–283.
Weidner, H. 1952. Insekten im Volkskunde und Kulturgeschichte. *Arbeitsgemeinschaft der Museen in Schleswig-Holstein. Niederschrift uber die Tagung der Arbeitsgemeinschaft am 28. und 29. Oktober 1950 im Heimatsmuseum in Rendsburg*, pp. 33–45.
WHO. 2013. Water-related diseases. (available at www.who.int/water_sanitation_health/diseases/malnutrition/en/). Accessed Feburary 2013.
WHO/FAO. 2010. INFOSAN Information Note No. 1/2010 – Biosecurity. *Biosecurity: An integrated approach to manage risk to human, animal and plant life and health*. Geneva, WHO.
WHO/FAO. 2012. Codex Alimentarius: international food standards. (available at www.codexalimentarius.org/). Accessed January 2013.
Wikipedia contributors. 2013. Maikäfersuppe. In *Wikipedia, The Free Encyclopedia*. Wikimedia Foundation, Inc. (available at http://de.wikipedia.org/wiki/Maik%C3%A4fersuppe). (last updated 23 February 2013).

Wilson, J.R.U., Ajuonu, O., Center, T.D., Hill, M.P., Julien, M.H., Katagira, F., Neuenschwander, P., Njoka, S.W., Ogwang, J., Reeder, R.H. & Van, T. 2007. The decline of water hyacinth on Lake Victoria was due to biological control by *Neochetina* spp. *Aquatic Botany*, 87(1): 90–93.

Winfree, R. 2010. The conservation and restoration of wild bees. *Annals of the New York Academy of Sciences*, 1195(1): 169–197.

Wolcott, G.N. 1933. *An economic entomology of the West Indies*. Puerto Rico, The Entomological Society of Puerto Rico.

Womeni, H.M., Linder, M., Tiencheu, B., Mbiapo, F.T., Villeneuve, P., Fanni, J. & Parmentier, M. 2009. Oils of insects and larvae consumed in Africa: potential sources of polyunsaturated fatty acids. *OCL – Oléagineux, Corps Gras, Lipides*, 16(4): 230–235.

Wood, J.R. & Looy, H. 2000. My ant is coming to dinner: culture, disgust, and dietary challenges. *Proteus*, 17(1): 52–56.

Xiaoming, C., Ying, F., Hong, Z. & Zhiyong, C. 2010. Review of the nuritive value of edible insects. *In* P.B. Durst, D.V. Johnson, R.L. Leslie. & K. Shono, eds. *Forest insects as food: humans bite back, proceedings of a workshop on Asia-Pacific resources and their potential for development*. Bangkok, FAO Regional Office for Asia and the Pacific.

Xing-Bao, J. & Kai-Ling, X. 1994. An index-catalogue of Chinese Tettigoniodea (Orthopteroidea: Grylloptera). *Journal of Orthoptera Research*, 3: 15–41.

Yen, A.L. 2002. Short-range endemism and Australian Psylloidea (Insecta: Hemiptera) in the genera *Glycaspis* and *Acizzia* (Psyllidae). *Invertebrate Systematics*, 16(4): 631–639.

Yen, A.L. 2005. Insects and other invertebrate foods of the Australian aborigines. *In* M.G. Paoletti, ed. *Ecological implications of minilivestock: potential of insects, rodents, frogs and snails*, pp. 367–388. New Hampshire, USA, Science Publishers.

Yen, A.L. 2009. Entomophagy and insect conservation: some thoughts for digestion. *Journal of Insect Conservation*, 13: 667–670.

Yen, A.L. 2010. Edible insects and other invertebrates in Australia: future prospects. *In* P.B. Durst, D.V. Johnson, R.L. Leslie. & K. Shono, eds. *Forest insects as food: humans bite back, proceedings of a workshop on Asia-Pacific resources and their potential for development*. pp. 65–84. Bangkok, FAO Regional Office for Asia and the Pacific.

Yen, A.L. 2012. Edible insects and management of country. *Ecological Management & Restoration*, 13(1): 97–99.

Yen, A.L., Hanboonsong, Y. & van Huis, A. 2013. The role of edible insects in human recreation and tourism. *In* R.H. Lemelin, ed. *The management of insects in recreation and tourism*, pp. 169–185. Cambrdige, Cambridge University Press.

Yhoung-aree, J. 2010. Edible insects in Thailand: nutritional values and health concerns. *In* P.B. Durst, D.V. Johnson, R.L. Leslie. & K. Shono, eds. *Forest insects as food: humans bite back, proceedings of a workshop on Asia-Pacific resources and their potential for development*, pp. 201–216.

Yhoung-Aree, J., Puwastien, P. & Attig, G.A. 1997. Edible insects in Thailand: an unconventional protein source? *Ecology of Food and Nutrition*, 36: 133–149.

Yhoung-Aree, J. & Viwatpanich, K. 2005. Edible insects in the Laos PDR, Myanmar, Thailand, and Vietnam. *In* M.G. Paoletti, ed. *Ecological implications of minilivestock*, pp. 415–440. New Hampshire, Science Publishers.

Yong-woo, L. 1999. Silk reeling and testing manual. *FAO Agricultural Services Bulletin*, 136.

Zagrobelny, M.A., Dreon, L., Gomiero, M.A.T., Marcazzan, G., Glaring, M.A., Linberg-Miller, B. & Paoletti, M.G. 2009. Toxic moths: a truly safe delicacy. *Journal of Journal of Ethnobiology*, 29: 64–76.

Zhang, C.X, Tang, X.D. & Cheng, J.A. 2008. The utilization and industrialization of insect resources in China. *Entomological research*, 38(1): 38–47.

Zoberi, M.H. 1973. Some edible mushrooms from Nigeria. *Niger. Field*, 38: 81–90.

Further reading

Aisi, C., Hudson, M. & Small, R. *How to ranch and collect insects in Papua New Guinea.* Cambridge, Cambridge University. (available at: www.geog.cam.ac.uk/research/projects/insectfarming/InsectManual.pdf).

Biomicicry Institute. 2012 (available at: www.biomimicryinstitute.org).

Bay Area Bug Eating Society. 2013 (available at: www.planetscott.com/babes/index.asp).

Cambridge University. 2012. Sustainable insect farming in Papua New Guinea. Cambridge, UK (available at: www.geog.cam.ac.uk/research/projects/insectfarming/publications.html).

Commonwealth Scientific and Industrial Research Organisation (CSIRO). 2012. Australian Dung Beetle Project (available at: www.csiro.au/Outcomes/Food-and-Agriculture/DungBeetles.aspx).

Cultural Entomology Digest. 2011. Insect articles (available at: www.insects.org/ced).

FAO. 2004. *Contribution des insects de la foret a la securite alimentaire. L'exemple des chenilles d'afrique centrale* (available at: www.fao.org/docrep/007/j3463f/j3463f00.htm).

FAO. 2013. Commission on Genetic Resources for Food and Agriculture – Micro-organisms and Invertebrates (available at: www.fao.org/nr/cgrfa/cthemes/cgrfa-micro-organisms/en/).

FAO. 2013. Overview on aquaculture and fish farm feeds (including some insect species) of World Fisheries and Aquaculture (available at: www.fao.org/fishery/topic/13538/en).

Girl Meets Bug. 2013. Edible Insects: The Eco-logical alternative (available at: www.girlmeetsbug.com/).

INRA, CIRAD, AFZ & FAO. 2013. Animal Feed Resources Information System of FAO (available at: www.feedipedia.com).

International Centre of Insect Physiology and Ecology. 2013. African insects Science for Food and Health (available at: www.icipe.org).

LINCAOCNET. 2013. Les insectes comestibles d'Afrique de L'Ouest et Centrale sur Internet (available at: http://gbif.africamuseum.be/lincaocnet).

Montana State University. *The Food Insects Newsletter.* Utah, USA (available at: www.foodinsectsnewsletter.org).

Network for Insect Collectors. 2013. The Network for Insect Collectors (available at: www.insectnet.com).

Oversease Development Institute. 2004. *Contribution of forest insects to food security and forest conservation: The example of caterpillars in Central Africa.* ODI Wildlife Policy Briefing, No. 3. London (available at: www.odi.org.uk/work/projects/03-05-bushmeat/wildlife_policy_briefs.htm).

Smallstock Food Strategies LLC. 2013. Smallstock (available at: www.smallstockfoods.com/about/).

South-South Cooperation. 2013. Partners in south-south cooperation (available at: www.southsouthcooperation.net/dvd/88-edible-insects-from-benin-to-costa-rica.html).

Venik. 2013. Dutch Insect Farmers Association (available at: http://venik.nl).

WUR. 2012. Insects as a sustainable feed ingredient in pig and poultry diets: a feasibility study (available at: http://www.wageningenur.nl/upload/ff5e933e-474b-4bd4-8842-fb67e6f51b61_234247%5B1%5D).

WUR. 2013. List of edible insects of the world. Wageningen, Wageningen University (available at: www.ent.wur.nl/UK/Edible+insects/Worldwide+species+list/).

**All internet links were checked on 18 March 2013.*

FAO TECHNICAL PAPERS

FAO FORESTRY PAPERS

1 Forest utilization contracts on public land, 1977 (E F S)
2 Planning forest roads and harvesting systems, 1977 (E F S)
3 World list of forestry schools, 1977 (E/F/S)
3 Rev.1 World list of forestry schools, 1981 (E/F/S)
3 Rev.2 World list of forestry schools, 1986 (E/F/S)
4/1 World pulp and paper demand, supply and trade – Vol. 1, 1977 (E F S)
4/2 World pulp and paper demand, supply and trade – Vol. 2, 1977 (E F S)
5 The marketing of tropical wood in South America, 1976 (E S)
6 National parks planning, 1976 (E F S)
7 Forestry for local community development, 1978 (Ar E F S)
8 Establishment techniques for forest plantations, 1978 (Ar C E* F S)
9 Wood chips – production, handling, transport, 1976 (C E S)
10/1 Assessment of logging costs from forest inventories in the tropics – 1. Principles and methodology, 1978 (E F S)
10/2 Assessment of logging costs from forest inventories in the tropics – 2. Data collection and calculations, 1978 (E F S)
11 Savanna afforestation in Africa, 1977 (E F)
12 China: forestry support for agriculture, 1978 (E)
13 Forest products prices 1960-1977, 1979 (E/F/S)
14 Mountain forest roads and harvesting, 1979 (E)
14 Rev.1 Logging and transport in steep terrain, 1985 (E)
15 AGRIS forestry – world catalogue of information and documentation services, 1979 (E/F/S)
16 China: integrated wood processing industries, 1979 (E F S)
17 Economic analysis of forestry projects, 1979 (E F S)
17 Sup.1 Economic analysis of forestry projects: case studies, 1979 (E S)
17 Sup.2 Economic analysis of forestry projects: readings, 1980 (C E)
18 Forest products prices 1960-1978, 1980 (E/F/S)
19/1 Pulping and paper-making properties of fast-growing plantation wood species – Vol. 1, 1980 (E)
19/2 Pulping and paper-making properties of fast-growing plantation wood species – Vol. 2, 1980 (E)
20 Forest tree improvement, 1985 (C E F S)
20/2 A guide to forest seed handling, 1985 (E S)
21 Impact on soils of fast-growing species in lowland humid tropics, 1980 (E F S)
22/1 Forest volume estimation and yield prediction – Vol. 1. Volume estimation, 1980 (C E F S)
22/2 Forest volume estimation and yield prediction – Vol. 2. Yield prediction, 1980 (C E F S)
23 Forest products prices 1961-1980, 1981 (E/F/S)
24 Cable logging systems, 1981 (C E)
25 Public forestry administrations in Latin America, 1981 (E)
26 Forestry and rural development, 1981 (E F S)
27 Manual of forest inventory, 1981 (E F)
28 Small and medium sawmills in developing countries, 1981 (E S)
29 World forest products, demand and supply 1990 and 2000, 1982 (E F S)
30 Tropical forest resources, 1982 (E F S)
31 Appropriate technology in forestry, 1982 (E)
32 Classification and definitions of forest products, 1982 (Ar/E/F/S)
33 Logging of mountain forests, 1982 (E F S)
34 Fruit-bearing forest trees, 1982 (E F S)
35 Forestry in China, 1982 (C E)
36 Basic technology in forest operations, 1982 (E F S)
37 Conservation and development of tropical forest resources, 1982 (E F S)
38 Forest products prices 1962-1981, 1982 (E/F/S)
39 Frame saw manual, 1982 (E)
40 Circular saw manual, 1983 (E)
41 Simple technologies for charcoal making, 1983 (E F S)
42 Fuelwood supplies in the developing countries, 1983 (Ar E F S)
43 Forest revenue systems in developing countries, 1983 (E F S)
44/1 Food and fruit-bearing forest species – 1. Examples from eastern Africa, 1983 (E F S)
44/2 Food and fruit-bearing forest species – 2. Examples from southeastern Asia, 1984 (E F S)
44/3 Food and fruit-bearing forest species – 3. Examples from Latin America, 1986 (E S)
45 Establishing pulp and paper mills, 1983 (E)
46 Forest products prices 1963-1982, 1983 (E/F/S)
47 Technical forestry education – design and implementation, 1984 (E F S)
48 Land evaluation for forestry, 1984 (C E F S)
49 Wood extraction with oxen and agricultural tractors, 1986 (E F S)
50 Changes in shifting cultivation in Africa, 1984 (E F)
50/1 Changes in shifting cultivation in Africa – seven case-studies, 1985 (E)
51/1 Studies on the volume and yield of tropical forest stands – 1. Dry forest formations, 1989 (E F)
52/1 Cost estimating in sawmilling industries: guidelines, 1984 (E)
52/2 Field manual on cost estimation in sawmilling industries, 1985 (E)
53 Intensive multiple-use forest management in Kerala, 1984 (E F S)
54 Planificación del desarrollo forestal, 1984 (S)
55 Intensive multiple-use forest management in the tropics, 1985 (E F S)
56 Breeding poplars for disease resistance, 1985 (E)
57 Coconut wood – Processing and use, 1985 (E S)
58 Sawdoctoring manual, 1985 (E S)
59 The ecological effects of eucalyptus, 1985 (C E F S)

60 Monitoring and evaluation of participatory forestry projects, 1985 (E F S)
61 Forest products prices 1965-1984, 1985 (E/F/S)
62 World list of institutions engaged in forestry and forest products research, 1985 (E/F/S)
63 Industrial charcoal making, 1985 (E)
64 Tree growing by rural people, 1985 (Ar E F S)
65 Forest legislation in selected African countries, 1986 (E F)
66 Forestry extension organization, 1986 (C E S)
67 Some medicinal forest plants of Africa and Latin America, 1986 (E)
68 Appropriate forest industries, 1986 (E)
69 Management of forest industries, 1986 (E)
70 Wildland fire management terminology, 1986 (E/F/S)
71 World compendium of forestry and forest products research institutions, 1986 (E/F/S)
72 Wood gas as engine fuel, 1986 (E S)
73 Forest products: world outlook projections 1985-2000, 1986 (E/F/S)
74 Guidelines for forestry information processing, 1986 (E)
75 Monitoring and evaluation of social forestry in India – an operational guide, 1986 (E)
76 Wood preservation manual, 1986 (E)
77 Databook on endangered tree and shrub species and provenances, 1986 (E)
78 Appropriate wood harvesting in plantation forests, 1987 (E)
79 Small-scale forest-based processing enterprises, 1987 (E F S)
80 Forestry extension methods, 1987 (E)
81 Guidelines for forest policy formulation, 1987 (C E)
82 Forest products prices 1967-1986, 1988 (E/F/S)
83 Trade in forest products: a study of the barriers faced by the developing countries, 1988 (E)
84 Forest products: World outlook projections – Product and country tables 1987-2000, 1988 (E/F/S)
85 Forestry extension curricula, 1988 (E/F/S)
86 Forestry policies in Europe, 1988 (E)
87 Small-scale harvesting operations of wood and non-wood forest products involving rural people, 1988 (E F S)
88 Management of tropical moist forests in Africa, 1989 (E F P)
89 Review of forest management systems of tropical Asia, 1989 (E)
90 Forestry and food security, 1989 (Ar E S)
91 Design manual on basic wood harvesting technology, 1989 (E F S) (Published only as FAO Training Series, No. 18)
92 Forestry policies in Europe – An analysis, 1989 (E)
93 Energy conservation in the mechanical forest industries, 1990 (E S)
94 Manual on sawmill operational maintenance, 1990 (E)
95 Forest products prices 1969-1988, 1990 (E/F/S)
96 Planning and managing forestry research: guidelines for managers, 1990 (E)
97 Non-wood forest products: the way ahead, 1991 (E S)
98 Timber plantations in the humid tropics of Africa, 1993 (E F)
99 Cost control in forest harvesting and road construction, 1992 (E)
100 Introduction to ergonomics in forestry in developing countries, 1992 (E F I)
101 Management and conservation of closed forests in tropical America, 1993 (E F P S)
102 Research management in forestry, 1992 (E F S)
103 Mixed and pure forest plantations in the tropics and subtropics, 1992 (E F S)
104 Forest products prices 1971-1990, 1992 (E/F/S)
105 Compendium of pulp and paper training and research institutions, 1992 (E)
106 Economic assessment of forestry project impacts, 1992 (E/F)
107 Conservation of genetic resources in tropical forest management – Principles and concepts, 1993 (E/F/S)
108 A decade of wood energy activities within the Nairobi Programme of Action, 1993 (E)
109 Directory of forestry research organizations, 1993 (E)
110 Proceedings of the Meeting of Experts on Forestry Research, 1993 (E/F/S) 111 Forestry policies in the Near East region – Analysis and synthesis, 1993 (E)
112 Forest resources assessment 1990 – Tropical countries, 1993 (E)
113 Ex situ storage of seeds, pollen and in vitro cultures of perennial woody plant species, 1993 (E)
114 Assessing forestry project impacts: issues and strategies, 1993 (E F S)
115 Forestry policies of selected countries in Asia and the Pacific, 1993 (E)
116 Les panneaux à base de bois, 1993 (F)
117 Mangrove forest management guidelines, 1994 (E)
118 Biotechnology in forest tree improvement, 1994 (E)
119 Number not assigned
120 Decline and dieback of trees and forests – A global overview, 1994 (E)
121 Ecology and rural education – Manual for rural teachers, 1995 (E S)
122 Readings in sustainable forest management, 1994 (E F S)
123 Forestry education – New trends and prospects, 1994 (E F S)
124 Forest resources assessment 1990 – Global synthesis, 1995 (E F S)
125 Forest products prices 1973-1992, 1995 (E F S)
126 Climate change, forests and forest management – An overview, 1995 (E F S)
127 Valuing forests: context, issues and guidelines, 1995 (E F S)
128 Forest resources assessment 1990 – Tropical forest plantation resources, 1995 (E)
129 Environmental impact assessment and environmental auditing in the pulp and paper industry, 1996 (E)
130 Forest resources assessment 1990 – Survey of tropical forest cover and study of change processes, 1996 (E)
131 Ecología y enseñanza rural – Nociones ambientales básicas para profesores rurales y extensionistas, 1996 (S)
132 Forestry policies of selected countries in Africa, 1996 (E/F)
133 Forest codes of practice – Contributing to environmentally sound forest operations, 1996 (E)

134 Estimating biomass and biomass change of tropical forests – A primer, 1997 (E)
135 Guidelines for the management of tropical forests – 1. The production of wood, 1998 (E S)
136 Managing forests as common property, 1998 (E)
137/1 Forestry policies in the Caribbean – Volume 1: Proceedings of the Expert Consultation, 1998 (E)
137/2 Forestry policies in the Caribbean – Volume 2: Reports of 28 selected countries and territories, 1998 (E)
138 FAO Meeting on Public Policies Affecting Forest Fires, 2001 (E F S)
139 Governance principles for concessions and contacts in public forests, 2003 (E F S)
140 Global Forest Resources Assessment 2000 – Main report, 2002 (E F S)
141 Forestry Outlook Study for Africa – Regional report: opportunities and challenges towards 2020, 2003 (Ar E F)
142 Cross-sectoral policy impacts between forestry and other sectors, 2003 (E F S)
143 Sustainable management of tropical forests in Central Africa – In search of excellence, 2003 (E F)
144 Climate change and the forest sector – Possible national and subnational legislation, 2004 (E)
145 Best practices for improving law compliance in the forest sector, 2005 (E F R S)
146 Microfinance and forest-based small-scale enterprises, 2005 (Ar E F S)
147 Global Forest Resources Assessment 2005 – Progress towards sustainable forest management, 2006 (E F S)
148 Tendencias y perspectivas del sector forestal en América Latina y el Caribe, 2006 (S)
149 Better forestry, less poverty – A practitioner's guide, 2006 (Ar E F S)
150 The new generation of watershed management programmes and projects, 2006 (E F S)
151 Fire management – Global assessment 2006, 2007 (E)
152 People, forests and trees in West and Central Asia – Outlook for 2020, 2007 (Ar E R)
153 The world's mangroves 1980–2005, 2007 (E)
154 Forests and energy – Key issues, 2008 (Ar C E F R S)
155 Forests and water, 2008 (E F S)
156 Global review of forest pests and diseases, 2009 (E)
157 Human-wildlife conflict in Africa – Causes, consequences and management strategies, 2009 (E F)
158 Fighting sand encroachment – Lessons from Mauritania, 2010 (E F)
159 Impact of the global forest industry on atmospheric greenhouse gases, 2010 (E)
160 Criteria and indicators for sustainable woodfuels, 2010 (E)
161 Developing effective forest policy – A guide, 2010 (E F S)
162 What woodfuels can do to mitigate climate change, 2010 (E)
163 Global Forest Resources Assessment 2010 - Main report (Ar C E F R S)
164 Guide to implementation of phytosanitary standards in forestry, 2011 (C E F R)
165 Reforming forest tenure – Issues, principles and process, 2011 (E S)
166 Community-based fire management – A review (E)
167 Wildlife in a changing climate (E)
168 Soil carbon monitoring using surveys and modelling – (E)
169 Global forest land-use change 1990–2005 (E F S)
170 Sustainable management of *Pinus radiata* plantations (E)

Ar – Arabic
C – Chinese
E – English
I – Italian
F – French
P – Portuguese
S – Spanish
R – Russian
Multil – Multilingual
* – Out of print

The FAO Technical Papers are available through the authorized FAO Sales Agents or directly from Sales and Marketing Group, FAO, Viale delle Terme di Caracalla, 00153 Rome, Italy.